The Red Helmet

UNITED STATES AIR FORCE

Thunderbirds

FLIGHT HELMETS

HISTORY - DEVELOPMENT - EVOLUTION

Jeffrey D. Guidry

EGP

ISBN 978-1-943492-76-3

ELM GROVE PUBLISHING
San Antonio, Texas, USA
www.elmgrovepublishing.com

Table of Contents

Dedication

This book is dedicated to the memory of

Major Stephen "Cajun" Del Bagno

who tragically lost his life on 04 April 2018 during a training flight over the Nevada Desert. Let us never forget "Cajun" and all the Thunderbirds who paid the ultimate sacrifice while serving this great nation as "Ambassadors in Blue."

30 April 1983 – 04 April 2018

Foreword

Since 1953, the United States Air Force Thunderbirds — America's "Ambassadors in Blue" — have performed for close to 300 million spectators at air shows in all 50 states and 57 foreign countries. Throughout those 66+ years, the Thunderbird legacy continues to grow, and has proven to be strong, and steeped in history and tradition. As a former Thunderbird Commander and Leader from 1977 to 1978, I am deeply proud of this legacy and know first hand the importance of preserving the rich heritage and traditions of the United States Air Force Aerial Demonstration Squadron.

The preservation of history can take on many forms, from the collection and conservation of significant artifacts, to monuments, and of course, the written text. While much has been published on the Thunderbirds, the author has chosen a subject previously unpublished and has produced a comprehensive reference work focusing on the history, development and evolution of the USAF Thunderbirds flight helmets. Following years of historical research, and through an artistic eye, the author provides the reader with a well balanced approach between authoritative and entertaining history. Several current and former Team members were interviewed, and their fascinating human interest stories are carefully woven throughout the book. These personal testimonies are inspirational, engaging, and give life and validation to this remarkable work.

This book is a genuine tribute to the importance of preserving the legacy of the USAF Thunderbirds, past, present and future, and pays homage to all the men and women who have had the honor and privilege of serving in this unique unit.

Dan Cherry

Brigadier General Edward "Dan" Cherry, USAF (Retired)
USAF Thunderbirds Commander / Leader 1977 - 1978
Author: *My Enemy, My Friend: A Story of Reconciliation from the Vietnam War*

Preface

The decision to write this book evolved slowly, over a lifelong passion for aviation. As a pilot, historian and collector of military flight equipment, a big part of my life has been devoted to my love affair with flying. I can't remember a time in my life when I didn't enjoy airplanes — from building plastic model kits as a young boy, earning my pilots license, building an experimental airplane, to owning and operating a small, private aviation museum.

Along this journey, I have evolved into a student of aeronautics and a purveyor of aviation memorabilia. I have amassed a respectable collection of military flight equipment from all U. S. service branches that includes flight helmets, flight suits, oxygen masks, anti-g suits as well as numerous ejection seats and cockpit instrument panels. At the centerpiece of this collection are authentic items from the U. S. Navy Blue Angels and the U. S. Air Force Thunderbirds.

My interest in the Thunderbirds dates back to October 19, 1975. The place was the former Chennault AFB in Lake Charles, LA. This was the first Thunderbird performance I attended — the first of many more to come. A beautiful fall Sunday as I recall. This was the second season (first full season) the Thunderbirds transitioned from the McDonnell Douglas F-4E "Phantom" to the Northrop T-38A "Talon" due to the 1973 fuel crisis. I still have vivid memories of this show, unfaded despite the passing of more than four decades.

As my interest grew in the Thunderbirds, I became particularly enamored with their flight helmets beginning in the early 1990s, and my research continues today. I learned early on that there was little, if any, documented information regarding the history, development and evolution of these beautiful helmets. Despite the vast electronic information repository we know as the Internet, details remained scarce. I also learned that the Thunderbirds had limited historical information to share on what turned out to be an interesting, albeit elusive story.

My initial goal was not to write a book but, after almost a quarter of a century of research, I felt a sense of motivation and compelled to delve into the unknown in an attempt to capture a story never before recorded. This journey has been incredibly educational, fascinating and frustrating, but above all, rewarding.

> **"I felt a sense of motivation, and compelled to delve into the unknown in an attempt to capture a story never before recorded."**
>
> Author

This endeavor has involved traveling thousands of miles and interviews of countless people — in person and by telephone — to insure this story was accurate and thorough. I included a human interest element by encouraging and recording the personal and anecdotal stories of several Thunderbird team members, both current and former, including a pilot from the first Thunderbirds[1] Aerial Demonstration Squadron established in 1953.

This book will take the reader on a journey that begins with the dawn of aviation, when pilots wore leather flight helmets, goggles and white silk scarves — daring young men in their flying machines. We witness the evolution of the flight helmet along the way to meet the needs of the jet age, when aircraft were flying faster and higher than anyone thought possible. As aircraft became more sophisticated, so did the flight helmets. In the interest of safety, cloth and leather flight helmets gave way to stronger materials, such as phenolic resins, fiberglass and high-tech Kevlar.

Over time, flight helmets evolved to meet specific safety and practical needs of the pilot as well as the mission. Although the Thunderbirds use standard, U. S. Air Force issued flight helmets, theirs go through additional aesthetic enhancements that include

---

[1] Originally named the Stardusters.

custom paint and flashy graphics, transforming these spherical canvasses into works of art, unlike any other in the Air Force.

We will explore several facilities, starting with the GENTEX plant where we witness the "birth" of a flight helmet, to Gibson & Barnes, the single source provider of custom flight helmets to the USAF Thunderbirds, and the Aircrew Flight Equipment Shop at Nellis AFB, Nevada.

"I am a student of history, not an expert. I have done my best to provide...the reader with a balanced approach between authoritative and entertaining history."

Author

I am a student of history, not an expert. My research has been meticulous, and I have endeavored to be as accurate as possible, citing sources and photo credits; however, in many cases, the various elements necessary to corroborate a story or prove a claim are all too often vague, or in some cases, non-existent. I have done my best to provide you the true story as I understand it, offering the reader a balanced approach between authoritative and entertaining history. This is a reference work based on documented history and the oral testimonies of many others. I have made a conscious effort to bring the reader the most comprehensive information available; however, as this book was being published, new and additional information has been discovered. This is not unique to this work. There comes a time to draw the line, and that time was now. As I continue my pursuit of history, discover additional data, conduct additional interviews, or uncover historical documents once believed lost forever, perhaps a second addition will be in order.

It is worth mentioning that this book is not a comprehensive, detailed history of U. S. Air Force flight helmets. These books already exist, and were written by authors with far more knowledge on the subject than myself. I have elected to provide the reader with an "entry-level introduction" of each of the flight helmets used by the Team to provide some meaningful context, with primary focus on the modifications, paint and custom graphics used by the Thunderbirds.

In addition to the historical merits, this book contains what I believe to be the most comprehensive collection of Thunderbirds flight helmet photos ever assembled in a single publication. These photos were gathered by pouring over thousands of images on the Internet, as well as through the generosity of current and former Thunderbird team members, aviation museums, and private collectors across the globe.

I hope that this book will be enjoyed by a diverse reader audience, including but not limited to Thunderbird fans, aviation enthusiasts, flight equipment collectors, and military aviation historians. It will appeal equally to those interested in detailed history, as well as those who enjoy the beauty of a lavish coffee table book.

Jeffrey D. Guidry

Introduction

Man's fascination with flying can be documented for over 2,000 years but, it wasn't until the Renaissance that we see notable advances in heavier than air flight. Fascinated by the phenomenon of flying, Leonardo Da Vinci studied and examined the flight characteristics of birds, and based on his observations and findings sketched plans for several flying machines. It would take another 400 years of research, and failed trials by many more inventors and aviation enthusiasts before man would eventually fly.

On December 17, 1903, the Wright brothers — Wilbur and Orville — made the first sustained, controlled, powered heavier-than-air manned flight at Kill Devil Hills, four miles south of Kitty Hawk, North Carolina.

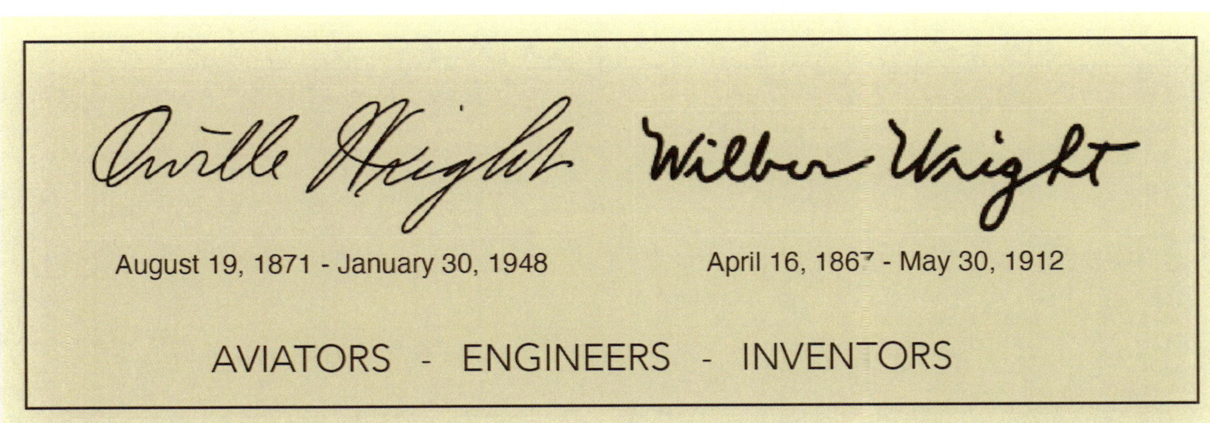

In a wood framed and fabric covered aircraft named the Wright Flyer 1, Orville recorded the first of several flights on that cold December day lasting a mere 12 seconds and covering a distance of 120 feet. Surprisingly, there were only five witnesses to this historic achievement. To this day many argue the aircraft is the greatest invention of the 20th century.

"The desire to fly is an idea handed down to us by our ancestors who looked enviously on the birds soaring freely through space on the infinite highway of the air."

Wilbur Wright

Since this dawn of aviation, through the end of World War II, aviators have recognized the need to wear helmets as primary protection from the harsh elements of the wind and cold. These early flight helmets were very basic and constructed of soft materials — cotton and nylon for summer, and leather for winter. For added warmth, some leather helmets were lined with wool, and later, electrically heated inserts.

Just 5 short years after that historic first flight,[1] Lt. Thomas E. Selfridge and Orville Wright were testing an aircraft recently purchased by the U.S. Army when the propeller separated, causing the aircraft to plummet uncontrollably to the ground. Orville survived, but Lt. Selfridge was killed when he was struck in the head by the separated prop. He was the first military aviator fatality in a heavier-than-air machine. His death emphasized the importance of protecting the pilots' heads.

In another accident report that appeared in *U.S. Army Aviation Digest* in 1913, two U.S. Army Signal Corps[2] pilots, Lieutenants Brereton and Chandler crashed their aircraft into the bay near Point Loma, California. Tragically, Lt. Chandler drowned and their aircraft was destroyed; however, Lt. Brereton survived with only minor injuries. The board of officers charged with the accident investigation stated: "Lietenant Brereton probably owes his escape from serious injury in part at least to his helmet as [he] received serious blows on his head." Unfortunately, we don't know much about the helmet Lt. Brereton was wearing at the time of the accident.

In those early years, little thought was given to the helmet providing the wearer with protection from injury. However, within 20 years of the first powered flight, efforts were underway in France, Italy and Germany to address the growing concern for head

[1] 17 September 1908

[2] Predecessor to the United Staes Air Force

protection. These "improvements" were typically in the form of padded leather, and afforded little, if any added protection.

"Lieutenant Brereton probably owes his escape from serious injury in part at least to his helmet as [he] received serious blows on his head."

With advances in aircraft design, the helmet began to take on additional functions, including the addition of ear phones to enable air-to-air and air-to-ground radio communications.

Helmets were also retrofitted with snaps to allow for the attachment of a mask for flights requiring supplemental oxygen at increasingly higher altitudes. Eye protection, in the form of simple metal-framed goggles with glass lenses were also essential, particularly in open-cockpit aircraft. Rigid visors were later introduced to protect pilots from the dangerous effects of wind blast during emergency egress.

As aircraft speeds increased, the need for protecting the pilot's head from inertial forces increased as well. Prior to the end of World War II, aeronautical laboratories in Germany were experimenting with a hard shell flight helmet borrowing from battlefield steel flak helmet designs.

From a 1912 issue of *Flight Magazine,* British inventor W.T. Warren is testing a protective flight helmet against a hangar wall, that he reported, "Will certainly attract considerable attention." The leather cap was padded with horsehair with a system of steel springs to intercept any impact, thus reducing the risk of injury.
(Photo credit: Wikimedia Commons)

In the United States, post war aircraft development at Muroc Field (now Edwards Air Force Base), such as the P-80 "Shooting Star" and the Bell XS-1 necessitated the need for an improved, hard shell flight helmet designed and built to provide the wearer with a much improved degree of head protection. Some test pilots, including USAF General Chuck Yeager, took matters into their own hands by modifying sports car racing helmets and football helmets, while waiting for the Air Force Aeronautical Lab at Wright Field (now Wright Patterson Air Force Base) to standardize a rigid "crash" helmet. Early U. S. Army Air Force jet squadrons faced a similar dilemma, and began using U. S. Army tanker helmets worn over their existing soft helmets. These "field" improvements, while innovative, were only a temporary fix to a serious and pressing matter of pilot safety.

In late 1947, the first standardized USAF production hard shell helmets came into use. The new helmets were constructed of molded nylon fabric and thermally set phenolic resins. These first production USAF hard shell helmets were designated the P-1, and had been inspired by Dr. Charles Lombard's (of Northrop Aviation) studies in helmet design at Wright Patterson AFB's Aero Laboratory. This new rigid flight helmet afforded the wearer improved impact resistance and reduced cockpit noise, but came at the cost of increased weight. The flight helmet would become one of many pieces of personal flight gear known as "life support equipment."

In that same year, while the jet age was still in its infancy, military aviation was hurtled into the future with the creation of the U. S. Air Force as a separate service. In just a few short years, the "jet age" was in full swing and the USAF realized a growing inventory of faster aircraft such as the F-84 "Thunderjet" and "Thunderstreak," F-86 "Saber," F-89 "Scorpion" and F-94 "Starfire."

Six years later, on May 25, 1953, the Air Force's official air demonstration team, designated the 3600th Air Demonstration Unit, was activated at Luke Air Force Base, Arizona. This unit eventually adopted the name "Thunderbirds," influenced in part by the strong Native American culture and folklore from the southwestern United States where Luke Air Force Base is located.

This is where the story begins. First, let's take a look at the Thunderbirds' mission, which states in part:

> **"The Thunderbirds' mission is to plan and present precision aerial maneuvers to exhibit the capabilities of modern, high-performance aircraft and the high degree of professional skill required to operate those aircraft.[3]"**

The Thunderbirds go to great lengths to insure that their aerial demonstrations are safely executed, in the air as well as on the ground. The precision routines, which are far from routine, are well rehearsed and choreographed with surgical precision.

3http://afthunderbirds.com/site/wp-content/uploads/2012/04/Thunderbirds-Fact-sheet-2013.pdf

Photo credit: USAF Thunderbirds

Known as "America's Ambassadors in Blue," the team strives to look as impressive on the ground as they do in the air. To accomplish this, the pilots are outfitted with fire retardant, hand tailored flight suits[4] from Gibson & Barnes[5] along with their signature custom painted flight helmets adorned with red, white and blue graphics, and the Thunderbirds trademark emblem at center stage![6]

My personal fascination with aviation dates back to my childhood, and has not diminished despite being well into my adult life. I was a typical boy, building plastic models of current jet aircraft, flying rubber band powered balsa wood airplanes and launching model rockets. I knew early on this was not a phase, or a passing hobby, but a genuine interest that would stand the test of time. I was not wrong. To this day, I still

[4] Through the years, the Thunderbirds have worn red, white, blue and grey flight suits.

[5] Formerly Flight Suits, Ltd., El Cajon, California.

[6] Gibson & Barnes, El Cajon, California.

gaze skyward whenever I hear an airplane flying overhead. Aviation had captured my imagination forever.

As a 5th grade elementary student, I closely followed the Apollo space program while many of my classmates were playing baseball and football. This was a choice, not a regret, as I was exhilarated to witness aviation history unfold before me. I was energized by my love affair with aviation and I was hungry for more.

In the summer of 1974, I cut out a $5.00 coupon from *FLYING* Magazine[7] for an introductory flight, redeemable at any participating Cessna Pilot Center. That 30-minute flight in a Cessna C-150 over south Louisiana changed my life forever.[8] I found a part-time job, learned to fly at 16 years of age and earned a private pilot's license before my driver's license. Unfortunately, defective color vision disqualified me from pursuing my dream job — flying for the U. S. military.

But, life goes on, and my passion for all things flying has not subsided. Instead, I became a student of aviation history, and eventually a collector of military aviation flight gear, instrument panels and ejection seats, which helped to fill the void left by forfeiting a career in aviation. As most collectors will agree, it is "the thrill of the hunt" that drives and motivates most of us, and I don't disagree; however, there is a very satisfying feeling that comes from preserving small pieces of aviation history and the stories that accompany these fascinating artifacts.

Over the years, I have amassed a respectable collection that I proudly exhibit in a small, private museum behind my home. This is a special place, where I entertain

---

7 The magazine was from an elementary school library where my father was Assistant Principal.

8 Paul Fournet Air Service, Lafayette, LA.

family, friends, guests and aviation enthusiasts with a common interest — a place where stories are shared and memories are made.

The flight gear in this collection is from the periods WWII, Korea, Vietnam, Desert Storm and present day, all of which are displayed on mannequins. Every service branch, as well as NASA and both U.S. flight demonstration teams are well represented. Without question, the centerpiece of this collection is a special section devoted to the USAF Thunderbirds that includes flight suits, flight helmets, uniforms, patches and miscellaneous memorabilia.

I am particularly drawn to the Thunderbirds flight helmets, and when I initially began my research for this book, was frustrated with the lack of historical documentation on this beautiful — *and critical* — piece of flight gear. Soon, my frustrations turned to motivation, and I have dedicated over 23 years to studying, documenting and photographing these elegant pieces of art at air shows, museums and private collections.

It is from that passion this book was born.

Join me on an historical and pictorial journey that takes an in depth examination of the development and evolution of the beautifully unique, and unquestionably stunning USAF Thunderbirds Flight Helmets for the past 66 years. From the team's humble beginning in 1953 to the present day and beyond, I explore deep behind the scenes to capture how these helmets are manufactured, built to exacting standards, artistically painted, customized and "earned" by each and every Thunderbird pilot.

1

Stardusters

On July 26, 1947 President Harry S. Truman signed into law the National Security Act which restructured the U.S. government's military and intelligence agencies. It also created three separate and distinct branches of the military: Army, Navy and a new Department of the Air Force. Shortly thereafter, on September 18, 1947, the United States Air Force was officially born and General Carl A. Spaatz would assume the role of Air Force Chief of Staff. By 1950, this massive restructuring effort was deemed a success.

As with any new organization, self-promotion was necessary to maintain the political as well as civilian support needed to fund aircraft, systems and programs required of a developing and growing air force. Senior leaders were motivated to show off their new service organization both at home and abroad, underscoring their efforts to recruit and retain qualified personnel.

In 1946 Admiral Chester Nimitz, Chief of Naval Operations also recognized the importance of self-promotion and had a vision to create a flight exhibition team in order to raise the public's interest in naval aviation and boost Navy morale. The name "Blue Angels" was suggested by Right Wing Pilot Lt. Maurice "Wick" Wickendoll after reading about the Blue Angel nightclub in The *New Yorker* magazine. A short three months following its birth, the Navy Flight Exhibition Team — the "Blue Angels" — performed its first flight demonstration on June 15, 1946, at their home base, Naval Air Station (NAS) Jacksonville,[1] Florida. The concept proved to be successful and very well received by the military as well as the public.

---

[1] Now based at NAS, Pensacola, FL

Not to be outdone, in May of 1953, the United States Air Force, 3600th Air Demonstration Team was created[2]. After just a few weeks of practice, the team became operational on June 1, 1953, and performed their first air show on June 16. Based at Luke Field, Arizona (now Luke AFB), the team was comprised of five pilots, all selected from the Air Force Advanced Flight Training School, also located at Luke.

Skyblazers Air Demonstration Team members (left to right)
Buck Pattillo, Dag Damewood, Bill Pattillo and Harry Evans. Photo credit: USAF

Major Richard "Smokey" Catledge, a training squadron commander at Luke, was selected as the team's first commander and would lead the team for the 1953 and 1954 show seasons. Maj. Catledge chose Captains Charles C. "Buck" and Cuthbert A. "Bill" Pattillo — identical twin brothers — for his left and right wingman as they had flown with the "Skyblazers" air demonstration team in Europe and were now instructors at

[2] This was the first dedicated USAF aerial demonstration team; Other USAF aerial demonstration teams pre-dated the Thunderbirds, but were collateral, part-time duties of the pilots assigned.

Luke. "Buck and Bill were highly experienced in this business and they in fact trained me," said Catledge. The slot pilot was Captain Bob Kanaga and a fifth pilot, Captain Robert "Mac" McCormick would be a backup pilot as well as perform solo maneuvers prior to the full diamond show.

"Buck and Bill were highly experienced in this business and they in fact trained me."

Major Richard "Smokey" Catledge
Stardusters / Thunderbirds Commander & Leader
1953 - 1954

Capt. McCormick reportedly said of the first public team performance, "They only gave us three weeks to get ready, and the first performance we did was in front of the Chief of Staff for the Air Force. There was a lot of pressure to get it right."[3]

Although the 3600th Air Demonstration Team was the official name, the decision was made to choose a less formal moniker, similar to their U. S. Navy counterparts, the "Blue Angels."

The Luke Field newspaper ran a contest to solicit ideas for the new team name. The contest ran for 30 days and as an incentive, the person submitting the winning name would receive a paid weekend at a Las Vegas hotel and a $50.00 U.S. Savings Bond.

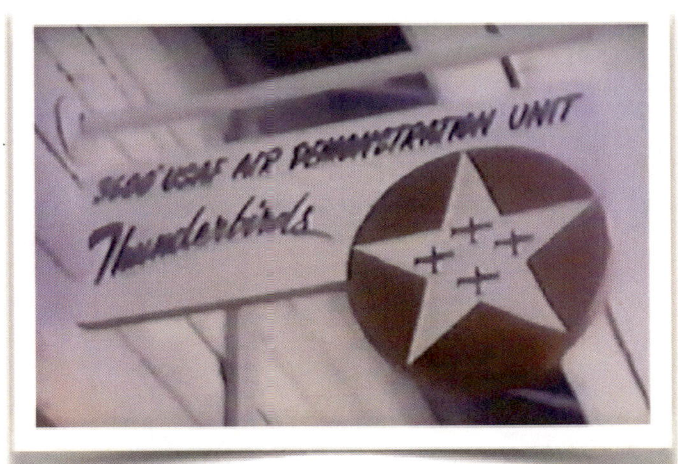

3 The Chief of Staff of the Air Force in 1953 was General Hoyt S. Vandenberg, who was attending his son's graduation from flight school at Williams AFB.

703

History Luke A.F.B. Ariz, DOCUMENT VI-3

January/June 1953

CORRECTED COPY | DESTROY ALL OTHERS

HEADQUARTERS
3600TH FLYING TRAINING WING (FIGHTER)
LUKE AIR FORCE BASE
Glendale, Arizona

GENERAL ORDER)
NUMBER 9)

23 May 1953

DESIGNATION, ORGANIZATION AND ASSIGNMENT OF 3600TH USAF AIR DEMONSTRATION FLIGHT (ACROJET)

1. Pursuant to authority contained in 4th Indorsement from Headquarters, Air Training Command, 8 April 1953 to letter, Headquarters, United States Air Force, File .001, Subject: "3600th USAF Air Demonstration Unit (Acrojet)", 11 December 1952, Air Training Command Regulation 20-4 and Message FMA 453E, Headquarters, Crew Training Air Force, dated 21 May 1953, the 3600th USAF Air Demonstration Flight (Acrojet) is designated and organized under the provisions of Air Training Command Table of Distribution Number CTN 105-63, with authorized strength of 8 officers and 26 airmen, effective 25 May 1953.

2. The 3600th USAF Air Demonstration Flight (Acrojet) is assigned to the 3600th Flying Training Group (Fighter), Luke Air Force Base, Arizona, effective with organization.

3. Personnel and equipment required for organization of above Table of Distribution Unit will be furnished from sources under the control of the Commanding General, 3600th Flying Training Wing (Fighter). Personnel will be assigned to unit and morning report will be initiated on date of organization.

4. Personnel will be attached to Headquarters Squadron Section, 3600th Flying Training Group (Fighter), for quarters, rations and administration effective with organization.

5. Applicable provisions of Air Force Regulation 20-52, as amended, Air Force Manual 181-5, Air Force Manual 171-6, Air Training Command Regulation 20-4, and Crew Training Air Force Regulation 20-1 will be complied with.

BY COMMAND OF BRIGADIER GENERAL BORN:

FRED N SCHMIDT
Major, USAF
Adjutant

OFFICIAL:

Fred N. Schmidt

FRED N SCHMIDT
Major, USAF
Adjutant

DISTRIBUTION: "B"
Headquarters, Air Training Command, Attn: DCS/Programming - 3 cys
Headquarters, Crew Training Air Force,
Attn: Manpower & Organization, DCS/Programming - 3 cys

The contest generated about 100 entries, with 25% including the word "Thunderbirds" in the name — the most common entry; however, despite the overwhelming popularity, the selection committee decided "Thunderbirds" was too regional, citing the prolific use of the word in and around Luke Field and Phoenix areas during the 1950s. Instead, the team announced SSgt. Fred Cesena as the winner with his entry, "Stardusters." History tells us this name was short lived, surviving less than 30 days.

Retired Col. Bob McCormick at the Eglin AFB 75th Anniversary Open House and Air Show, 2010.
U.S. Air Force photo by Staff Sgt. Richard Rose

Word of the new team name soon reached Lt. Gen. Bob Harper at Training Command Headquarters, Scott Air Force Base, Illinois, and Maj. Catledge learned by way of telegram Harper's dislike for the name "Stardusters." Instead, Harper instructed the team to choose a name based on the regional folklore of the southwestern United States, suggesting "Thunderbirds" had a nice ring to it.

Shortly after the new name was finalized, Captain Bob McCormick was tasked with sketching the now famous Thunderbird logo. Capt. McCormick went into the caves in the desert near Luke AFB to research the Native-American deity known as the Thunderbird. According to folklore, the Thunderbird made thunder by flapping its wings, and had the ability to hurl lightning from its place in the sky. Capt. McCormick did a pencil rubbing on the cave walls to use as the design for the Thunderbirds' patch.

He started the design using a white disc with an attached white scroll below the disc, all edged in black. The most prominent feature of the emblem is a rendering of the ancient Indian god, Thunderbird, with a yellow beak and eye, surmounted by a red disc charged with a white five-point star, all above an arched inscription "Thunderbirds" in black script. Red, white and blue were taken from the Native American design, but also represented the national colors. The Thunderbird lacks the lightning bolts of war and the olive branches of peace since neither fit into the squadron's mission, which is to display modern, high-performance aircraft. The Thunderbirds patch is a symbol of the rich tradition of the entire unit which has been worn by every member of the Thunderbirds since the creation of the team with little modification over the years.

F-84G "Thunderjet"
3600th AIR DEMONSTRATION TEAM "THUNDERBIRDS"
LUKE AFB, 1953
Prints available from aviationgraphic.com

Original Thunderbirds Team, August 1953.
Photo Credit: USAF

Perhaps coincidentally, in addition to the selection of Thunderbirds as the official team name, the first fighter airplanes assigned to the group by the Air Force were nicknamed the "Thunderjet". The Air Force's frontline fighter of the day, the Republic F-84G, would serve the team for the first two seasons.

To set the team's jets apart, Major Catledge wanted a distinctive paint job. On day one, and before selecting any pilots, he knew he wanted a red, white and blue color scheme with blue stars on a white background on the vertical stabilizer (tail).

On one of the first pilot meetings, the team kicked around how to paint the aircraft. Without reaching a consensus, Maj. Catledge gave Capt. McCormick the task of putting some ideas on a small F-84 model. "Mac" took the model home and returned the following day with the red scallops painted on the nose, fuel tip-tanks, and blue stars on the white vertical and horizontal stabilizers. They liked the paint job, but thought the scallops flowed too much.

Photos from Author's personal collection.

McCormick took another model home, made some design changes and came back the next day with the color scheme that was adopted. The aircraft were painted without delay.

Initially the team wore standard issue, U. S. Air Force, K-2, olive drab flight suits. The Republic Aviation Company — manufacturer of the F-84 — was eager to do something for the Air Force demonstration team flying their aircraft. According to Maj. Catledge, they asked Republic to have some distinctive uniforms made for the team, but the uniforms they made were said to be a "terrible shade of blue" and didn't fit! Maj. Catledge's wife, Norma, solved the problem, suggesting to dye their standard issue Air Force flying suits black. She was given all of the flight suits, placed them in a washing machine and dyed them black, which became the flying suits the Thunderbirds wore for several years.

Unfortunately the stories and details surrounding the first Thunderbirds patch and flight suits are more revealing than their flight helmets which remains somewhat of a mystery. However, it is clear that the helmet designs closely mimic the aircraft paint schemes particularly the vertical and horizontal stabilizers which provides us some interesting clues. As the saying goes, "A picture is worth a thousand words."

And so, the story begins.

Staff Sergeant Jacques "Jack" Moran
USAF Stardusters / Thunderbirds 1953 - 1955
Group Operations Officer

In 2016, while researching the USAF Thunderbirds, I stumbled across a story on the Internet detailing the retirement of a longtime Mississippi Power & Light employee. Upon reading the story, I learned that the retiree — Jack Moran — was one of the founding members of the original USAF Stardusters / Thunderbirds team in 1953.

Jack Moran had a colorful military career, serving as Group Operations Officer for the original U. S. Air Force Thunderbirds aerial flight demonstration team. The group was formed in 1953 in part by Major Dick Catledge, the man who just happened to be Moran's commanding officer at the time.

"I joined the Air Force in 1951 as a typist because I took typing in high school," Moran said. "I was stationed in San Antonio as a Staff Sergeant when Major Catledge said he had been asked to be the commanding officer of a new unit. He asked me if I wanted to go with him to work on it. I asked what we would be doing, and he told me I'd find out when we got there."

So Moran was off to Luke Air Force base in Glendale, AZ, with 29 others to get the squadron up and running.

Photos courtesy of Jack Moran

"I worked on the team's travel orders, we got the airplanes painted, and I even got the enlisted team's first jump suits at a local J.C. Penney's department store. I got everybody's sizes, bought the suits, and took them to the base laundry to have the patches and names sewn on."

"I was 23 years old so it was a lot of fun," Moran recalled. "Two of the original four pilots were twins.[4] Imagine that. Only 10 of the original 29 are still alive today."

[4] Buck and Bill Pattillo

With any new undertaking there's bound to be some dicey moments. Moran says he saw a few of those during his time with the Thunderbirds.

"I was always interested in flying," he said. "I got to go on numerous missions across the country and overseas. On one trip, the team was set to perform in Guatemala and I was on the support plane with several others. The plane developed engine trouble over a third-world country and we had to make an emergency landing on a small runway in the jungle. After that, I always made sure I was wearing my parachute. Even though it was awkward at first, we ended up staying the night there and the locals got us groceries."

"There was another time when the public affairs officer decided to take the president of a small country up in his jet for a ride. The problem was everyone in town was lined up on the runway to watch. Our guy asked the president what he wanted to do and the president said, 'Take off!' The crowd loved it."

Anecdotes aside, Moran says it was an honor to be part of the original Thunderbirds team and even at 85 years old[5], he's still heavily involved in the Thunderbirds Alumni Association.

"I only had two jobs in my life," Moran added. "Being in the Air Force and working with the Thunderbirds, and being at Mississippi Power. I loved them both."

Through some crafty investigative work, I located Mr. Moran's son, also an employee of Mississippi Power & Light. I successfully made contact with him in the summer of 2018, and shortly thereafter, I received a phone call from a gentleman who identified himself as "Jack Moran, former Thunderbird!"

It was evident Mr. Moran was very proud to be a part of this inaugural team. He said he admired Maj. Catledge and held him in high regard, and would have worked for him

[5] In 2016

anywhere, regardless of the assignment. He didn't hesitate to accept the job offer from Maj. Catledge despite not knowing what it was or where it was located!

I listened intently to several stories of this first team, the respect he had for Maj. Catledge and how "lucky" he was to be part of the inaugural team. His memory and zeal for this rare opportunity was still vibrant and detailed.

He told me about arriving at Luke AFB and that one of the first orders of business was to purchase unique uniforms for the enlisted team. The original uniforms were sourced from the local J.C. Penney department store! He also arranged for the team's first patch to be sewn to the front, and "Thunderbirds" embroidered across the back.

He also shared that Capt. Bob "Mac" McCormick (spare pilot) jokingly wishing ill fate on one of the show pilots so he could take their place!

Mr. Moran laughed when he told me that the identical twin Pattillo brothers - Buck and Bill - were quite the pranksters, often confusing their team mates on their true identity, much to the chagrin of Maj. Catledge!

U.S. Thunderbirds "Zippo" Lighter
Personal Collection of Jack Moran

In the Winter of 2019 I managed to arrange for a follow-up telephone interview with Mr. Moran who still seemed excited to kindly share a few more memories with that first Thunderbirds team.

Mr. Moran reiterated his fondness for Maj. Catledge, and described his management style as "Laid back." He commented several times in our conversation that he was the best boss he ever had — military or civilian life.

One of Mr. Moran's primary duties with the Thunderbirds was writing up the team's travel orders for Maj. Catledge's approval and signature. The bond and trust was so strong between the two men that Maj. Catledge suggested Mr. Moran sign his (Catledge) name to the orders instead of waiting for approval.

Mr. Moran recalled asking Maj. Catledge if he could travel with the team on occasion to air show sites, to which Maj. Catledge replied, "Don't you know how to write your name?" With that, Mr. Moran began adding his name to travel orders, signing for Maj. Catledge, and enjoying some time on the road with the team.

Aerobats To Leave On Tour

Special to The Republic

LUKE AIR FORCE BASE—The Thunderbirds, official USAF aerobatic demonstration team based here, will depart Wednesday on a 35-day goodwill tour of South America.

The tour will cover an estimated 15,000 miles and will include 11 South American cities. All types of air force planes will be demonstrated for the first time in many of the cities.

The Thunderbirds will fly at the following places: Mexico City; Managua, Nicaragua; Panama Canal; Lima, Peru; Santiago, Chile; Buenos Aires, Argentina; Montevideo, Uruguay; Rio de Janeiro, Brazil; Maracay, Venezuela; Cuidad Trujillo, Dominican Republic, and Havana, Cuba.

Maj. Richard C. Catledge will lead the Thunderbird diamond formation of F-84 Thunderjets. Capts. Cuthbert and William Pattillo will fly the wing positions and Capt. Robert S. McCormick will hold the slot position.

Lt. A. D. Brown will be the spare pilot and maintenance officer. Accompanying the Thunderbirds will be 13 maintenance men, a flight surgeon, public relations officer, and Capt. Alphonse D. Quinones, interpreter.

We talked briefly about those first Thunderbird flight helmets, and I asked Mr. Moran if he had any recollection about how the design was selected. He believed Capt. Bob "Mac" McCormick was responsible as he had designed the aircraft paint scheme and agreed that the helmets closely resembled the design selected for the F-84s.

Speaking of Capt. McCormick, Mr Moran told a story of working one day in the squadron offices along with Maj. Catledge, Capt. Bill Pattillo and Capt. Buck Pattillo. Mr. Moran had a very simple transistor radio, and he selected a country station for their listening pleasure as this was his favorite genre of music. Without warning, Capt.

McCormick reportedly stood up, walked over to Mr. Moran's desk and changed the station, announcing, "I don't like that damn stuff." Capt. Bill Pattillo, a southerner from Atlanta, GA said, "I like it. Let's take a vote." Capt. McCormick, who Mr. Moran described as "Kind of a jackass" said, "Moran doesn't get a vote!" Capt. Bill Pattillo stepped in and said, "Of course he gets a vote, he owns the damn radio!" In the end, they continued listening to country music!

From the private, personal collection of
SSgt. Jack Moran

This was a different era. This was a unique squadron — the only one of its kind. Military protocol was somewhat relaxed. Many of the team members were from the South, including Mr. Moran, which added to the genteel atmosphere of the team. Prior to his military service, Mr. Moran worked on a dairy farm in southern Mississippi, a geographical region also known, in some circles, for illicitly produced, high proof, distilled spirits — Moonshine! It is "rumored" that certain team members, who will remain anonymous, had a "taste" for the bootleg hootch and would place "orders" when Mr. Moran went home on military leave. We shared a good laugh!

Mr. Moran said that after Maj. Catledge left the team, and Capt. Broughton took the helm, the atmosphere and dynamics were very different. The gentle, easy going management style of Maj. Catledge had come to an end. Capt. Call was a military academy graduate, and Mr. Moran remembers, "There was a lot of saluting going on in the squadron." He took a couple of weeks leave to consider his options, during which time he was offered a job with Mississippi Power and Light. He accepted the offer and would not to re-enlist in the Air Force.

When Capt. Broughton took over, "There was a lot of saluting going on in the squadron."

SSgt. Jack Moran

Thank you, Sir, for your service to our great nation, and sharing your wonderful stories of that inaugural Stardusters / Thunderbirds flight demonstration team!

This image below is from the back cover of *AIR Training*, October 1953, Volume 3, Number 3. *AIR Training* is the official internal informational publication of the Air Training Command, published monthly by Headquarters, ATRC, Scott AFB, Illinois. (From Author's personal collection)

Original USAF Thunderbirds patch, ca 1953, F-84G. Courtesy Peter Vonk. Background photo: USAF

2

P-1-Series Flight Helmet

The first production hard shell flight helmet issued by the Army Air Force was designated P-1, making its debut on 12 September 1947, 6 days before the birth of the U.S. Air Force as a separate service. The early P-1 helmets were a big improvement over the cloth and leather helmets that afforded the wearer with no impact protection.

Helmet, Flying, Protective, Type P-1A, United States Air Force, Smithsonian Air & Space Museum.

The first P-1s were manufactured by the Paramount Rubber Company, with a subsequent contract award to the Switlik Parachute Company, Trenton, New Jersey in 1948.

The P-1 helmet was constructed of pressure and heat molded laminated cotton duck fabric that had been saturated with phenolic-resin (essentially similar to "Bakelite" materials), and produced in only one size. The P-1 shell was issued in white color with a decal of the Army Air Force winged star on a blue circular field. Some pilots painted their helmets with colorful squadron markings or to suit the wearer's personal taste.

ADJUSTING
THE
P-1A
HELMET
1 HEAD BAND LACING
5 SHELL LACING KNOT
2 NECK STRAP CORD LACING
3 SLING CORD LOOP
4 RECEIVER MOUNTING ANCHOR LACING
1 Head Size is adjusted with the head band lacing.
2 Earphone Pad Pressure is adjusted with the neck strap cord lacing.
3 Head Depth is adjusted with the center sling cord loop. In no case should the shell be less than one-half inch from the top of the head. In all helmets with nylon slings, this loop should be tied, then the ends of the cord doubled back through the webbing loops and again tied. This permits absorbing higher shock loads with less travel of the shell.
4 Lateral Earphone Adjustment is made by removing the receiver mounting anchor lacing at the bottom anchor point and relocating in new holes punched in the mounting flange.
5 Shell Lacing, which suspends the shell to the sling a safe distance from the head, must be kept tight at all times. Little safety is afforded if the lacing is loose. The ends of the lacing on back of the shell holds the communication cord in place. All lacing ends should be securely tied with square knots and any excess lengths of cord ends tied off with double bow knots.
HELMETS PREVENT INJURIES
HOTCH
Department of The Air Force ● Inspector General, USAF ● Directorate of Flight Safety Research ● Norton Air Force Base, San Bernardino, Cal. No. 14 Aug. 1950
Prepared for Aero Medical Laboratory Headquarters Air Materiel Command.

Over the next few years, the P-1 would go through some slight modifications, including offering two sizes - small and large, improved leather oxygen mask attach tabs, a fixed chin strap, and improved ear cup retention.

A P-1 helmet, thus modified, was re-designated as a P-1A and later re-designated P-1B[1]. Interestingly enough, neither Switlik nor Paramount Rubber won any production contracts for the updated helmet. Instead contracts were awarded to at least three other companies, Lite Manufacturing Company, General Textile Mills (GENTEX), and Bassons Industries Corporation

Identification tags from the inside shell of a P-1A, Lite Manufacturing Company, Inc., and P-1B General Textile Mills, Inc.
(Author's Personal Collection)

The First USAF Thunderbirds Flight Helmet

There is very little historical data available on the first USAF Thunderbirds flight helmet with the exception of a few rare photos, and fewer fading memories of past team members. So much institutional knowledge has been lost to the passing of time, and the passing of too many of those early pilots and ground crews. However, we can draw some logical conclusions based on these photos, the time period they would have been issued, and a few context clues.

[1] Identical to the P-1A; Nomenclature change only when reissued.

Flashback

Stardusters

Photo courtesy Joe Patrick

Shortly after its birth at the end of the Korean War, USAF's 3600th Aerial Demonstration Flight was dubbed "Stardusters." The name, SSgt. Fred Cesena's entry in a contest at Luke AFB, Ariz., was short-lived. Within a month of the team's first official performance—June 16, 1953, in honor of USAF Chief of Staff Gen. Hoyt S. Vandenberg's impending retirement—the name was changed to the now-familiar "Thunderbirds." The Stardusters are shown with General Vandenberg at Williams AFB, Ariz.: (counterclockwise, from left) Capt. Buck Pattillo, Maj. Dick Catledge, Capt. Bob McCormick (alternate), Capt. Bob Kanaga, and Capt. Bill Pattillo (Buck's twin).

On the previous page is a copy of an article from "Air Force Magazine" (July 1994) showing some of the earliest photos of the very first Thunderbirds flight helmets. These helmets are clearly P-1A / P-1B helmets that were common to the time period, and standard issue with USAF flight crews, with the exception of the custom paint and graphics. Unfortunately, this first helmet "artist" remains a mystery.

As we examine other available photos, we can see a unique pattern of 3 different sized stars, forming 2 inverted "V" patterns on the top and rear of the helmet shell. I have been unsuccessful in determining the significance of these "V" patterns, sizes or number of stars. There are far more questions than answers.

I am also making an educated assumption that the stars were painted, as was the standard at the time. It wasn't until much later that vinyl graphics became common practice due to convenience and improved quality.

Wall Mural, USAF Thunderbirds Museum, Nellis AFB, Nevada (Author's photo)

Despite the relative low quality of these photos, it is evident that the helmet was not disassembled prior to receiving the custom paint job. This is most evident by the cord that weaves horizontally in and out of the helmet shell (to secure the internal suspension harness). The cord is painted red in those areas covered by the "scallop" design.

This rare USAF photo (above) provides a nice view of the rear of the helmet featuring the "Inverted Double-V" star pattern. The photo on the left is a reproduction from the author's personal collection.

(Left) A very early, undated photo of the 3600th Air Demonstration Team with their black dyed (previously olive) USAF-issue K-2 flight suits and P-1A or P-1B flight helmets.

Pictured (left to right): Capt. C.C. Pattillo, Capt. R. L. Kanaga, Capt. C. A. Pattillo, and Maj. R. C. Catledge.

(Photo credit: USAF)

Capt. C. A. "Bill" Pattillo standing on his F-84G "Thunderjet" holding his P-1A or P-1B flight helmet. (Photo credit: USAF)

I have searched the Internet, collaborated with other flight helmet collectors, and poured over countless books about the Thunderbirds only to fall short of uncovering any detailed information on this first flight helmet, with the exception of the photographs I have collected and included in this chapter.

(Above) Maj. R. C. Catledge, Capt. C.C. Pattillo, Capt. C. A. Pattillo, and Capt. R. L. Kanaga posing with their F-84G "Thunderjet".
(Photo credit: USAF)

According to the book "DIAMOND IN THE SKY - A Pictorial History of the United States Air Force Thunderbirds" by C.A. Knotts & Pete Moore, the P-1A / P-1B flight helmet was used between 1953 and 1955 when the team flew the F-84G "Thunderjet" and F-84F "Thunderstreak". As beautiful as these F-84-era P-1A and P-1B flight helmets were, they were the only ones in the team's history to not use the Thunderbird emblem. That would soon change.

As suggested in Chapter 1, there is a strong indication that this original flight helmet design, featuring the red scallops and blue stars on a white field, closely follow the paint scheme on the F-84G "Thunderjet," particularly on the vertical and horizontal stabilizers, as shown below. We know from previous stories that Capt. McCormick possessed some artistic talent since he was charged with the design of the Thunderbirds patch as well as the F-84G paint scheme. It would not be surprising to learn that Capt. McCormick also had a hand at designing these original P-1 flight helmets, which seemingly have his "fingerprints" all over them!

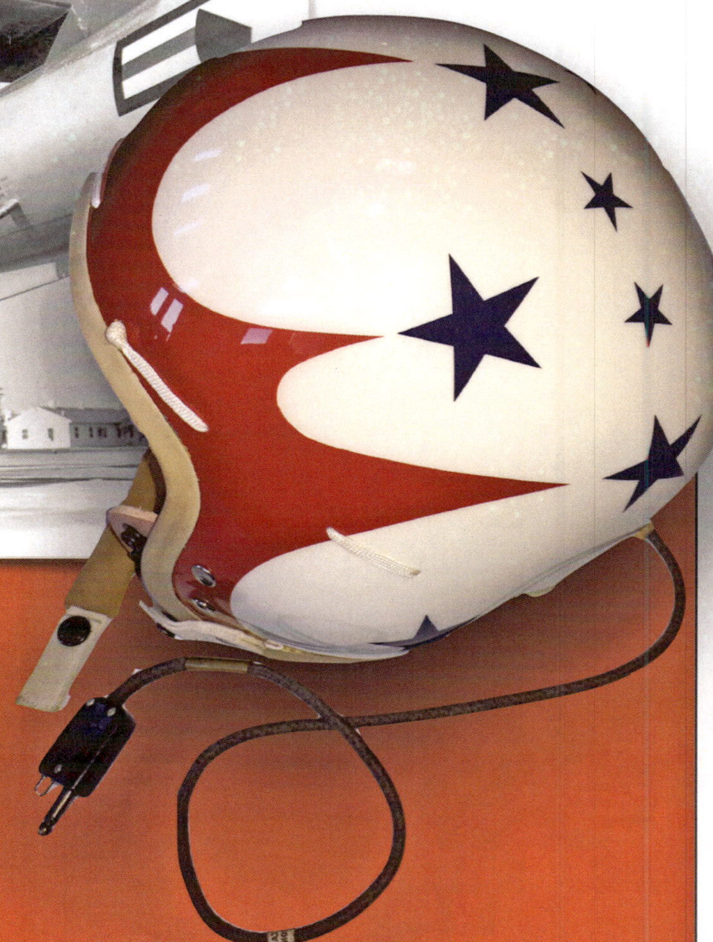

Capt. C. C. "Buck" Pattillo F-84G, Luke AFB, June 1953, photo credit: USAF (above).

Right: P-1 flight helmet from author's collection.

The interior of these helmets featured a very basic nylon and leather suspension harness that was secured to the helmet via a nylon cord woven through a series of holes drilled through the helmet shell. The nylon cord (right photo) could also be used to adjust the size of the suspension harness to insure a better fit for the pilot. The suspension harness also featured a leather "crown" in the center that essentially supported the helmet on the pilot's head. The crown was stenciled in white with the "winged star" and "U. S. AIR FORCE."

These slings and harnesses, manufactured by General Textile Mills, Inc. (GENTEX) in this example, were a big improvement over their leather and fabric predecessors, but still left a lot to be desired when it came to comfort following extended wear from long missions. It would take several years before additional research and technology provided better fitting and comfort solutions.

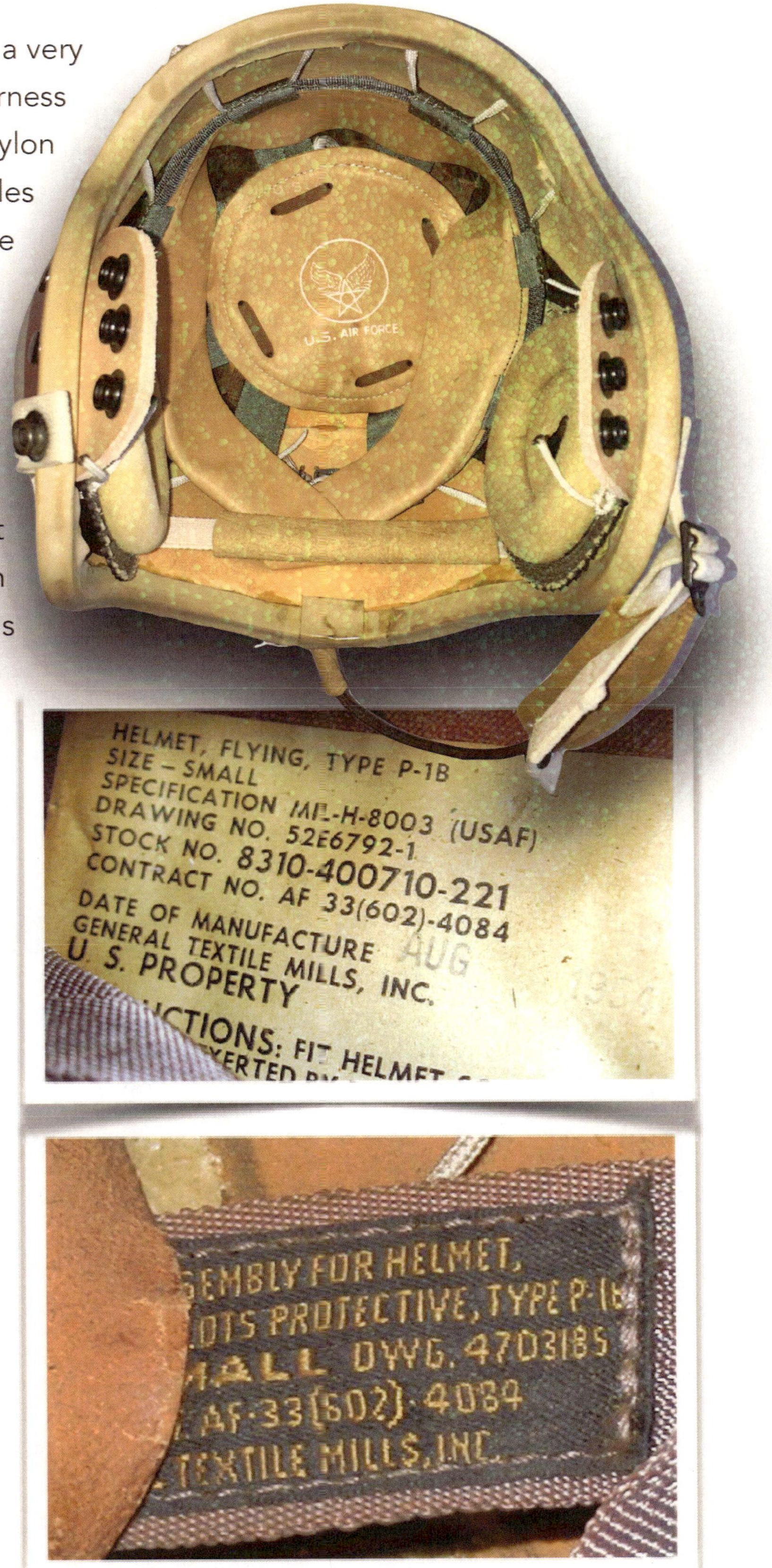

P-1 Flight Helmet from Author's Collection.

All images on this page are screenshots from a USAF Thunderbirds history video, and feature the original Stardusters P-1 flight helmet.

Precision Formation Maneuvers To Be Demonstrated At Dayton Air Show Soon

DAYTON, O. — Precision formation maneuvers used daily by USAF fighter pilots in combat will be seen by spectators at the National Aircraft Show, Dayton Municipal Airport, Dayton, September 5, 6 and 7 when five crack pilots of the Air Force's new precision formation flying team, the Thunderbirds, put their Republic F-84G's through their paces.

Precision flying performed by the team will demonstrate the speed and maneuverability of our nation's first-line fighter aircraft. Another purpose of the team is to emphasize the level of skill attained by Air Force fighter pilots in their flying training.

The precision demonstration, lasting approximately 15 minutes, will start with a take-off entry into a diamond formation as the four silver, red, white and blue USAF F-84's leap off the runway. Their maneuvers will progress from a diamond formation through rolls, loops and clover leaf turns to a grand finale of the "bomb burst."

It is in this thrilling spectacle that the four jets climb straight up in box formation and at about 5,000 altitude split up, roll and then head earthward as fragments of a shell bursting, meeting simultaneously from four directions at more than 500 miles per hour.

The Journal Tribune
Marysville, Ohio
Thursday, August, 27, 1953

The Republic F-84, used for close ground support operations in combat, has a speed in excess of 600 miles per hour, with a combat radius of over 850 miles. The jet engine produces over 5,600 pounds of thrust to speed them through the air. In combat, these F-84's would carry two 1,000 pound bombs, 32 five inch rockets and six .50 calibre machine guns with which to destroy an enemy.

The team's formation includes four positions, referred to as a diamond formation. The front point is the lead man. Slightly behind him and to his right and left are the wing men. The man flying the rear point position in the diamond is referred to as the "slot" man, and it is considered probably one of the most difficult positions to fly.

Stationed at Luke Air Force Base, Glendale, Ariz., the Thunderbirds are instructors in Luke's fighter combat crew training program as part of the Crew Training Air Force. The team was activated June 1, 1953.

Flying the lead position in the team, and the man who plans all the maneuvers, is Major Richard C. Catledge, 32 year old native of Memphis, Tenn., who flew 23 missions from North Africa during WW II before being shot down and taken prisoner.

In the wing positions are "hard to tell apart" Georgia twins, Captains C. A. "Bill" and C. C. "Buck" Pattillo. Both were members of the "Skyblazers" precision team in Germany from 1949 to 1952. Both flew F-51 Mustangs in World War II, where "Bill" was shot down over Germany and taken prisoner.

Slot position on the team is flown by a 28-year-old Kansan, Captain Robert L. Kanaga, who has tallied up 93 missions in the F-51 and 12 in the Lockheed F-80 Shooting Star in Korea.

Alternate on the team to fill and of the positions is Captain Robert S. McCormick, 28, from Wisconsin. He flew the right wing position in the Sabre Dancers, an F-86 demonstration team. Capt. McCormick has chalked uo 100 missions in Korea where he flew a North American F-86.

Photo above is from *LIFE* Magazine, August 3, 1953 (Author's collection).

The 1953 Stardusters Team
(Photo Credit: U.S. Air Force)

Jet Pilots Display Air Maneuvers Here

Five jet pilots of the 3600th Air Demonstration team put on an air show for the benefit of Sheppard airmen and civilians last Thursday. The five F-84 G Thunderjets put on a precision flying demonstration from the new north-south runway.

Demonstrating maneuvers used by jet pilots in combat, the pilots went through the barrel roll, cloverleafs, loops, and bomb bursts.

The team, base at Luke AFB, Arizona, was headed by Major Richard C. Catledge. Other members of the team were twin captains, C. A. (Bill) Patillo and C. C. (Buck) Patillo. Rounding out the crew were Captain Robert L. Kanaga and Capt. Robert S. McCormick, alternative.

Now on a nation-wide exhibition tour, the team left for James Connally AFB in Waco following the Sheppard visit.

The Sheppard Senator
Wednesday, July 8, 1953
Sheppard Air Force Base

Lieutenant General Charles C. "Buck" Pattillo, USAF (Retired) USAF Stardusters / Thunderbirds Left Wing 1953 -1954

USAF Photos.

Lieutenant General Charles Curtis "Buck" Pattillo, USAF (Retired) was born on June 3, 1924 in Atlanta, GA, and is the twin brother of Major General Cuthbert A. "Bill" Pattillo, USAF (Retired).[2]

Both Pattillo brothers enlisted in the U. S. Army Air Force as aviation cadets, where they earned their wings and commissions as second lieutenants in 1944.

(Above) Lt. Charles C. "Buck" Pattillo (left) and Lt. Cuthbert A. "Bill" Pattillo (right) stand in front of Buck's P-51 Mustang, "Little Rebel." (Photo credit: starduststudios.com)
(Below) Lt. Charles C. "Buck" Pattillo with the "Skyblazers. (Photo credit: aerobaticteams.net)

[2] Died February 20, 2014

They joined the 352nd Fighter Group, 8th Air Force, where they flew P-51 Mustangs in the European Theater of Operations.

Their careers continued on a parallel path when they were stationed in Germany and instrumental in organizing and flying with the "Skyblazers," the U. S. Air Forces in Europe aerial demonstration team.

Upon returning to the United States, once again, the Pattillo twins were stationed together at Luke Field, Arizona, where in 1953, they both flew with the USAF Stardusters, known today as the Thunderbirds.

(Left to Right) 1st Lt. A.D. Brown, Capt. Bob McCormick, Capt. Bill Pattillo, Maj. Richard Catledge, Capt. Buck Pattillo.
Photo Credit: ThunderbirdsAlumni.com

In the winter of 2018, I learned that Lieutenant General Charles C. "Buck" Pattillo, USAF (Retired) — one of the original Stardusters / Thunderbirds pilots — was living unassisted in Virginia at the age of 94. With considerable research, and a little luck, I was able to locate one of his sons, Colonel Charles C. Pattillo, Jr., USAF (Retired). I wrote a letter to Col. Pattillo explaining my book research and strong desire to interview his father, who is the sole surviving pilot of the original Stardusters / Thunderbirds team.

1953 USAF Stardusters
Photo Credit USAF

Much to my delight, Col. Pattillo promptly replied and we began a meaningful dialogue. I was elated to learn that Lt. Gen. Pattillo was receptive and agreeable to an interview, but he expressed concern through his son that his fading memory might not produce much meaningful information. Col. Pattillo advised me that his father would contact me directly!

Imagine my utter surprise when my phone rang on a lazy Sunday afternoon in early January 2019 from a number I didn't recognize. The caller said, "This is Buck Pattillo... is this Jeff?" I was so surprised by the call that I was having difficulty containing my excitement and maintaining my composure!

Despite being 94 years young, Gen. Pattillo was still sharp, articulate and spoke with the hypnotizing drawl of a southern gentleman. He was quick to apologize that he may not have much to contribute due to a fading memory, but felt compelled to offer whatever memories he could muster in the interest of Thunderbirds history.

It was evident early in the conversation that Gen. Pattillo remains passionate about his association with the Thunderbirds, and came across as eager to share his unique story.

I was a bit surprised when Gen. Pattillo kicked off our interview by asking the first question, "Do you know about the Stardusters?" I replied, "Yes, sir...yes, sir I most certainly do!" What followed was more of a casual conversation than an interview, and to be quite honest, I was a little star struck! I stumbled out of my chair and fumbled around for a pen and paper so I could scribble down the notes of our conversation. Following the call, I immediately sat down to transcribe my notes that resembled a doctor's prescription — practically unreadable! I think my wife described me as a "giddy schoolgirl!"

"Do you know about the Stardusters?"

Lieutenant General C. C. "Buck" Pattillo, USAF (Retired)
USAF Stardusters / Thunderbirds
Left Wing 1953 - 1954

Not surprising, we quickly began talking about flight helmets, and I found It interesting to learn that when the Stardusters were initially formed, the Team flew with standard USAF issued P-1 flight helmets, painted in the standard white. Gen. Pattillo recalled that it was sometime later that the red scallops and blue stars were added, but couldn't remember any details surrounding the customization[3]. We discussed the similarity of the helmet design to that of the aircraft, in particular, the vertical stabilizer and rudder, and

[3] The Team had custom painted flight helmets by their first official performance on 16 June 1953.

he agreed that it was likely the work of Capt. Bob "Mac" McCormick and 1st. Lieutenant A. D. "Brownie" Brown, the Maintenance Officer and spare pilot.

He added that in the beginning, the team also wore standard issue USAF olive drab flight suits, as discussed in Chapter 1. He described them as "old and smelly!"

Gen. Pattillo chuckled as he told me his version of the familiar story about the Stardusters' early beginnings, how they got their name, and quickly changed it to Thunderbirds.

I was reminded that the name Stardusters was selected via a contest sponsored by the Luke AFB newspaper; however, when Lt. Gen. Bob Harper at Training Command Headquarters, Scott Air Force Base, Illinois learned of the selection, he sent word to Major Catledge — the team commander / leader — suggesting "Thunderbirds" would be a more appropriate name given their geographical location which was richly steeped in Native American culture.

K-2 flight suit from Author's Collection. Standard issue P-1 Flight Helmet courtesy of the Smithsonian Air & Space Museum.

Because advertising for the first few air shows had already been published in local newspapers, those shows were billed and flown as the USAF Stardusters. When this news made its way to Lt. Gen. Harper, he sent word back to Maj. Catledge advising him, "I don't think you understood what I meant!" The situation was quickly explained, and the Stardusters were soon performing as the USAF Thunderbirds.

Gen. Pattillo expressed his pleasure with efforts to continue recording the history of the Thunderbirds. He mentioned having several footlockers in his basement that may contain letters, paperwork and perhaps photos of his time with the Stardusters and Thunderbirds, offering to search through them and forward any items relative to my research. I felt a strong sense of privilege and honor.

I took the opportunity to share the photo on the left with Gen. Pattillo to positively identify the pilot, as I have seen it listed as both "Buck" and "Bill" Pattillo from different sources. Gen. Pattillo was quick to respond, "This one is of Bill - I was better looking!

"This one is of Bill – I was better looking!"

Lt. Gen. Charles C. "Buck" Pattillo

I received a follow-up phone call from Gen. Pattillo in late February 2019; unfortunately, the foot locker scavenger hunt failed to produce any fruit! However, while we were talking, Gen. Pattillo took me on a "guided tour" of what he called his "Me Room" — a special place within his home with memorabilia and photographs from his stellar, life-long career in the Air Force. As he looked at all of the photos hanging silently on the wall, he described to me what he saw in each of the framed memories. He specifically sought out photos of the Thunderbirds, but didn't have any that showed the Team with their iconic flight helmets.

Gen. Pattillo told me that his son, Chuck, shared some photos I sent to him of a restored reproduction of the first Thunderbird P-1 flight helmet. He said that the blue stars and the red scallops were very similar to the stars and red scallops that adorned their Republic F-84G "Thunderjets" he flew in 1953. He continued, "At that time, you could have whatever style of helmet you wanted...If you wanted a pink helmet, you could have a pink helmet." However, "Most squadrons would agree on a design, and all the pilots in the squadron would get the same design. This was the case with the Thunderbirds." He added, "I believe the flight helmets have changed over the years, but I think the design has remained fairly consistent."

Gen. Pattillo apologized for not having more to offer, but I felt like I had just won the lottery!

> **"He was our Superman, strong to the end, and we were blessed to have him for almost 95 years."**
>
> Col. Chuck Pattillo, USAF (Retired)

Gen. Pattillo mentioned that he would be attending the 2019 Joint Base Andrews air show as a guest of the Thunderbirds, and suggested this might be a good opportunity to meet. He also agreed to take a photograph with my fully restored P-1 flight helmet in Thunderbird colors for the book. This was an opportunity not to be missed!

As I was making the leisurely drive to Joint Base Andrews in early May, I received an email from Gen. Pattillo's son, Chuck, informing me that his father had been hospitalized, and the likelihood of his air show attendance was highly unlikely. Sadly, less than two weeks later, Gen. Pattillo passed away, marking the end of an era. Chuck, speaking about his father told me, "He was our Superman, strong to the end, and we were blessed to have him for almost 95 years."

Rest in peace, Sir.

IN FORMATION ON GROUND stands the team: Major Catledge, Captains "Buck" *(left)* and "Bill" Petillo, Captain Kanaga. Rear: Captain McCormick.

DANGEROUS TEACHING

Skillful group of Air Force jet pilots takes hazardous way of inspiring cadets

The dramatic precision flying shown on the opposite page is considered an educational maneuver, not a stunt, by the U.S. Air Force. From observing it, say officers, cadets can see how the very best pilots fly and will be inspired to perfect their own flight skills. To give this kind of spectacular instruction the first official jet team has been set up at Luke Air Force Base in Arizona. The leader is Major Richard C. Catledge; the two wing positions flying behind him are Captains "Buck" and "Bill" Petillo, identical twins of Atlanta, Ga., and the rear "slot" is flown by Captain Robert L. Kanaga.

Doing precision acrobatics like barrel rolling and power diving at speeds of more than 500 mph is a highly hazardous job. Only the leader sees where he is going. The others have to follow blindly by taking bearings on his tail and wing tips. So tight are team's formations, exhaust of the lead jet frequently blackens the tail of the plane flying rear slot. The team never goes up in the air without its fifth member, Captain Robert S. McCormick, who is ready to fill in immediately at any position should one of the four regulars have to drop out.

This photo and article is from the August 3, 1953 issue of *LIFE* Magazine.

I was alerted to this article by Patricia Pattillo Landis, daughter of the late Maj. Gen. C. A. "Bill" Pattillo and niece of Lt. Gen. C. C. "Buck" Pattillo, who sent me photos from the magazine on display in her home. I was able to secure my own copy through an eBay auction.

This is yet another great historical photo of this first team just weeks following their inception.

The title of the article, *DANGEROUS TEACHING* is without a doubt an attention getter, but an unlikely approach today.

Lt. Gen. Charles C. "Buck" Pattillo

03 June 1924 — 20 May 2019

3

P-3 Series Flight Helmet

As aircraft speeds became increasingly faster, pilots continued to be plagued by helmet retention during emergency egress. By 1948 development was underway of an externally mounted, rigid visor that would prove to be more stable than the goggles used with the P-1-series flight helmets. Beginning 26 April 1951, visor modification kits were ready for distribution and retrofitting to existing P-1A and P-1B helmets.

The modification kit included a new rigid side-latching external visor assembly that replaced the B-8 rubber goggles. Goggles couldn't withstand the wind blast from an ejection and would be blown off of the helmet. The new visor assembly was a big improvement in oxygen mask retention and eye protection. The visor had two secure positions, full up or full down.

P-3 series flight helmet. (Photo credit unknown)

P-1A and P-1B helmets modified with the new visor assembly were designated P-3. It took almost two years from the time the first modification kits were installed until the first factory-made P-3 helmets became available. Production contracts were awarded to the Selby Shoe Company, General Textile Mills and Bassons Industries Corporation. The P-3 helmet became mandated for use in all high-speed, high performance USAF jet fighters and bombers.

As the mainstream USAF transitioned from the P-1A and P-1B flight helmets to the P-3, so did the Thunderbirds.[1]

In the USAF photo on the right, we see a P-3 helmet with the external visor attached. This is clearly an F-84G aircraft (in service between May 1953 and the spring of 1955), and the pilot appears to be Bill Creech, Right Wing for the 1954 - 1955 show seasons. From this information, we can deduce that this photo is likely from 1954, and provides evidence of when the Thunderbirds began mak ng the transition from the P-1 series helmets to the P-3 series helmets.

While difficult to be certain, the red "scallop" pattern appears unchanged from the P-1A and P-1Bs; however, some design changes would soon be noticed.

On the next page is an undated USAF photo of Jack Broughton, who commanded the Thunderbirds from October 1954 to February 1957. Broughton led the team through the transition from the straight-wing F-84G to the swept-wing F-84F, and then to become the world's first supersonic acrobatic team in the F-100C "Super Sabre." Based on the ejection seat in the photo, and the canopy style, this aircraft was the F-84F "Thunderstreak" which entered service in 1955.

1 The P-2 flight helmet — not mentioned in this study — was a concept only and never went into commercial production.

The photos on this page shows one of the modifications made to the P-1 series helmets to convert them to P-3 helmets. The highlighted area is an aluminum bracket (one on each side) for attachment and function of the external visor.

USAF Photo

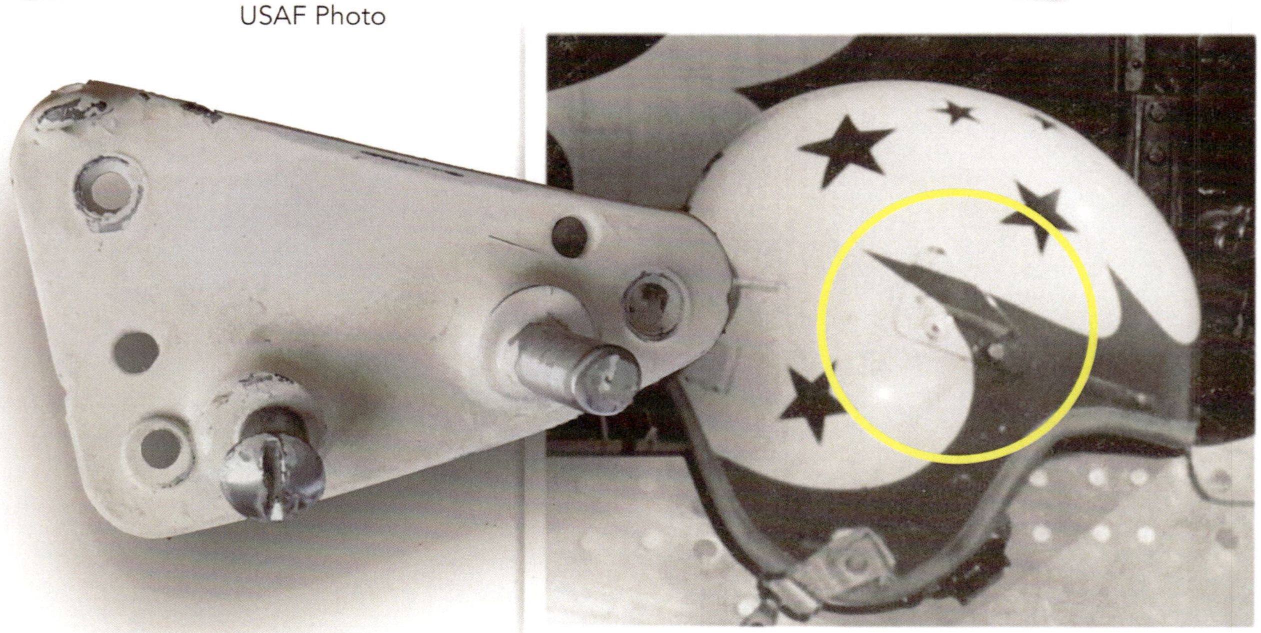

(Above) Late 1954 period posing with the F-84G "Thunderjet" are Jack Broughton, Leader, Bill Creech, Right Wing, Ed "Lucky" Palmgren, Slot and Burt Spaulding, Left Wing. (Photo credit: McGuire AFB Photo Archives)

In the example on this page, from late 1954, it appears that one of the paint schemes involved routing the red scallops around the aluminum visor bracket, rather than "through" the bracket as seen in the previous pages.

The decision to route the red scallops around the bracket make sense, as it is an "obstruction" and very difficult to tape and paint. As a matter of personal opinion, the scallop lines do not flow as nicely as they do when routed through the brackets.

Similar to the P-1-series flight helmets, the P-3 continued to use the same nylon and leather suspension harnesses, offering no improved comfort solutions to the wearer. Fortunately, that would soon change.

As a point of discussion and interest, there are clearly some variations on the red scallops, as seen on the previous pages, particularly in the manner that they transition through or around the aluminum visor bracket.

The photo on the previous page (credit: USAF) is from around 1955 - 1956 based on the F-84F "Thunderstreak" aircraft. The embroidered names and aircraft positions on the pilots' flight suits also provide some useful clues; however, based on available historical data of all Thunderbirds officers who served from 1953 to present day, the names don't quite align with the photo. The pilots are likely Jack Broughton, Bill Creech, Burt Spaulding and possibly Ed "Lucky" Palmgren.

All four pilots look very proud as they "present" their beautiful, custom painted P-3 flight helmets for the camera! They certainly look happy, and understandably so!

If you examine the photo closely, you will notice that the star pattern on the helmet held by the pilot on the far left of the photo is different from the other three helmets! The classic "Inverted V" star pattern discussed earlier in this chapter are clearly consistent with the helmets in the 12-o'clock, 3-o'clock and 6-o'clock positions in the photo, while the helmet in the 9-o'clock position follows a distinctly different pattern!

This is the only such discrepancy I have encountered in examining the limited number of available Thunderbirds P-1 and P-3 flight helmets. Was this a mistake, or an oversight, or perhaps a prototype? Unfortunately for collectors and historians, this question is likely to remain unanswered, and we may never know the story on this anomaly!

Sometime around 1955, the Thunderbirds would introduce a significant design change to their P-3 flight helmets.

Another version of the P-3 style helmet appears as a simple line drawing in the book, "DIAMOND IN THE SKY - A Pictorial History of the United States Air Force Thunderbirds." This is the first reference we see that includes the classic Thunderbirds emblem, and the only time it was displayed on the front of the helmet. Subsequent helmet upgrades and design changes moved the emblem to the rear of the helmet where it has remained for decades.

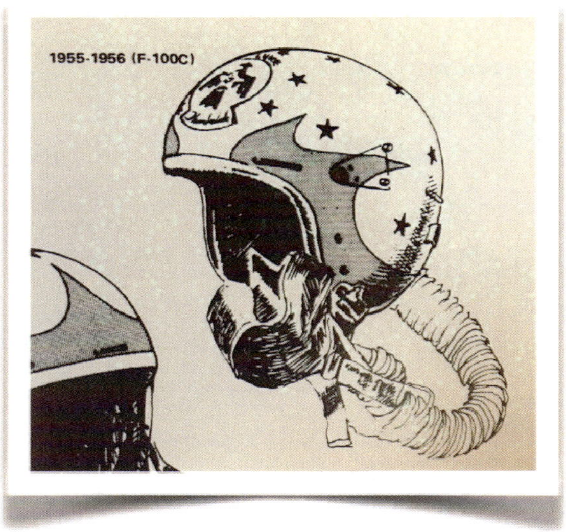

According to the reference on the photo, this helmet was only used for the 1955-1956 show seasons (end of F-84F and beginning of F-100C), making this specimen very rare. It is interesting to note that even in this line drawing, the modifications are clearly illustrated, showing the visor mounting brackets; however, without the visor attached. It is unclear why the P-3 helmets were flown with the external visor removed.

As an artist and historian, I am intrigued with these designs and periodic changes and curious as to who was behind the decision making process. Who were these artists? Were they from the team, or commercial service providers? Regardless, the designers and painters showed beautiful artistic talent and flair on their spherical canvasses.

I was fortunate to locate a P-3 Thunderbirds helmet from a private collector I have come to know through a shared passion for the Team.[2] He shared the photos on the next page with me, but could provide no history or provenance on the helmet, only that it was purchased at an estate sale. A rare example, indeed.

[2] John Slezak, North Carolina

LUKE
Thunderbirds

LUKE

The previous photos are beautiful examples of a very unique and very rare helmet. The Thunderbird image appears to be hand painted, along with the blue stars. The 4 aircraft in the center of the star are clearly the "F" model of the F-84, the distinctive swept wing "Thunderstreak." The "LUKE" banner above the Thunderbird emblem is also consistent with aircraft markings of the same period.

Thunderbirds support aircraft, C-119 "Flying Boxcar" 1955.
Photo Credit: Airman Magazine

Although not a complete example, it tells an interesting story of the evolution of the scallop design and star pattern. The most obvious change is on the front of the helmet where the center red scallop "point" was removed to make way for the now famous Thunderbird emblem.

Upon closer examination, we can confirm the star pattern has also been modified. On the P-1 helmet (left), there were stars at the tips of each of the 5 scallop points; however, on the P-3 (below), the stars were moved to the inside of the scallops, where they reside to this day. The large inverted "V" star pattern remained, but with different sized stars, and the smaller inverted "V" pattern was shortened.

These P-3 flight helmets were originally painted in the same pattern as their P-1 predecessors. Later examples reveal the scallop and star patterns were modified to incorporate the Thunderbirds emblem on the front of the helmet, as shown in the photo below.

This poor quality image is of Capt. Ed "Lucky" Palmgren, slot pilot, is believed to be from 1955.

The helmet is a P-3 series featuring the revised red scallop design and introduction of the first Thunderbird emblem on the front.

Although not the best quality, the image on the next page is of Capt. Ed "Lucky" Palmgren, and provides a nice drawing of the rarely seen P-3 flight helmet. This image and the images on the previous page are from http://www.vvmf.org/Wall-of-Faces/39326/EDWIN-D-PALMGREN.

VIETNAM VETERANS MEMORIAL FUND

FOUNDERS OF THE WALL

Capt. Edwin "Lucky" Palmgren twice served with the USAF Thunderbirds, first from 1954 - 1955 as the solo pilot, and again from 1963 - 1964 as the Commander / Leader. He served in World War II, Korea and Vietnam and was posthumously awarded the grade of Colonel.

While flying an F-111 Aardvark mission over Vietnam (tail number 66-0024, call sign "Tailbone 78"), Col. Palmgren and his crew member , Capt. David L. Cooley (a U.S. Navy pilot on exchange with the USAF) never returned to their base at Tahkli Royal Thai Air Base, Thailand. He was listed as Missing In Action (MIA) on 22 April 1968 and later declared Killed In Action (KIA) on 26 November 1973. Despite search efforts, the crew and aircraft were never recovered.

Col. Palmgren is remembered and memorialized by the Vietnam Veterans Memorial Fund — founders of the Vietnam Memorial Wall — as well as the Wall of Faces (https://www.vvmf.org/Wall-of-Faces/). Col. Palmgren can be found on panel 51E, line 032 of the Vietnam Memorial Wall in Washington, DC.

USAF Photo

The 1955 USAF Thunderbirds with a 1955 Ford Thunderbird! Pictured from left to right are Jack Broughton, Commander / Leader; "Mac" McCormick, Solo; Burt Spaulding, Left Wing; Bill Creech, Right Wing; Ed Palmgren, Slot, unknown far right. (USAF Photo)

So, where have all of the P-1 and P-3 flight helmets gone? Other than the P-3 shell belonging to a private collector shown earlier in this chapter, I have been unable to locate any other examples in museums or personal collections. First, most pilots of this era were only issued one flight helmet — this includes the Thunderbirds. Given what we know about the makeup of these first two teams, we are only talking about plus or minus 10 helmets — a minuscule number considering the thousands produced for all of the Air Force for the same period. Like many rare artifacts, some may have been destroyed, others repainted and reissued, or perhaps a few tucked away in barns, storage lockers or basement closets, just waiting to be re-discovered!

4

TOPTEX, Model T-3

The P-Series flight helmets continued to evolve with the introduction of a lighter fiberglass shell, improved communications and a modified external visor latching mechanism; however there are no records or photographs to suggest the Thunderbirds used this version, designated the P-4, and later improvements, the P-4A and P-4B.

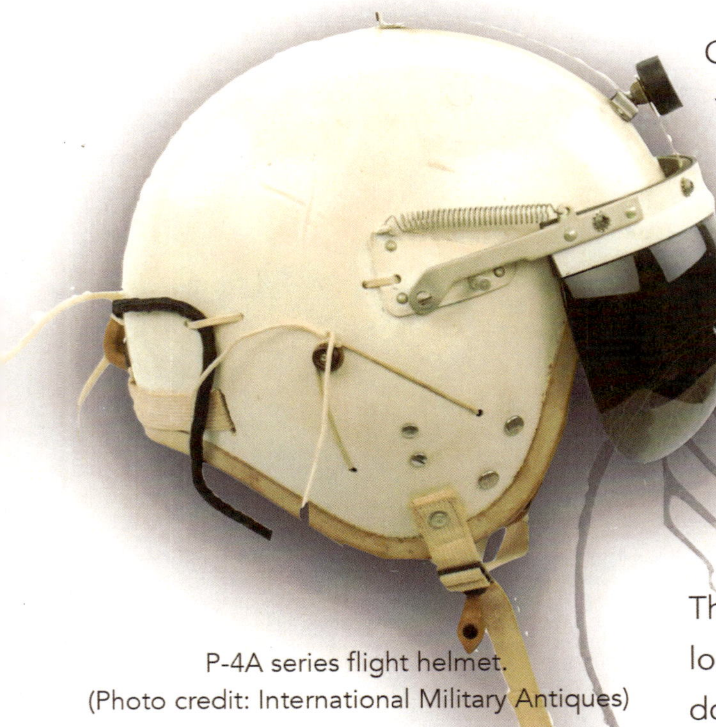

P-4A series flight helmet.
(Photo credit: International Military Antiques)

One might assume this would be a natural transition, but the Thunderbirds instead selected the TOPTEX flight helmet, produced by Protection Inc.,[1] a Division of Mine Safety Appliance Company (MSA). The TOPTEX T-3 was similar to the P-4 but typically favored by Test pilots since the middle of the 1950s through the 1970s, as well as in some select USAF squadrons.

This helmet has a unique external visor locking mechanism with central track and does not utilize a chin strap. In the TOPTEX early configuration, the external visor was not present and the oxygen mask was secured to the snaps on leather tabs, replaced in the later versions by Hardman[2] external oxygen mask "Christmas tree bayonet[3]" receivers. External boom mounted microphone, clear visor or customized mask receivers were in some cases also installed.

[1] Acquired by General Textile Mills (GENTEX) in 1969.

[2] Hardman Tool & Engineering Company.

[3] Named because the ratcheted bayonets resembled a Christmas tree.

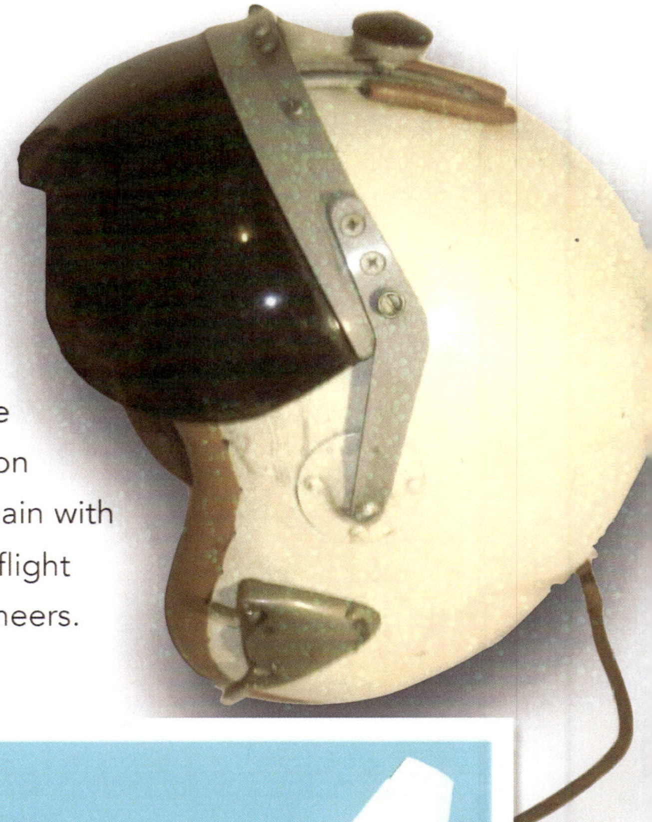

The TOPTEX was developed by Dr. Charles Lombard[4] who based his design off of the P-1 helmet. The crown of the shell was slightly flattened and featured a custom fashioned, individualized fit inner liner that was shaped to the specific wearer's cranial conformation. The helmet design did not include a chin strap. Instead, the oxygen masked served to secure the helmet to the wearer's head. The TOPTEX aviation helmet soon gained a reputation that would remain with it through the 1970s as the preferred protective flight helmet of professional flight test pilots and engineers.

Top: Mine Safety Appliance, TOPTEX Flight Helmet. (Author's Personal Collection)
Below: The "Original Seven" Mercury Astronauts - all with TOPTEX Flight Helmets (NASA Photo).

[4] Dr. Lombard's studies in helmet design at Wright Patterson AFB's Aero Laboratory.

The TOPTEX was so favored by USAF test pilots, that Edwards AFB wrote a Sole Source of Supply Justification[5] for these custom flight helmets citing pilot safety.

EDWARDS AIR FORCE BASE, Edwards California

Subject: Sole Source of Supply Justification

1. The following information is submitted for further justification on the Sole Source of Supply pertaining to Purchase Requests indicated.

a. HELMET, TOPTEX, CRASH PROTECTIVE. Subject helmet is a very strong, custom-fitted, shock-absorbing crash helmet which was designed by Protection, Inc. for use primarily by test aircrews associated with crash landings, emergency escape, and cockpit buffeting. With experimental test aircraft there is a greater possibility of emergency conditions than with production model aircraft. Actual stability test programs conducted at this Center require that a test aircraft be flown to the extent of its capability. This means that the pilot will be buffeted around the cockpit and suffer terrific blows on the head. Due to the close proximity of the canopy, canopy rails, instrument panel and control column, it is essential that the pilot's head will be protected in order to prevent head injury and unconsciousness.

(1) The use of these helmets has saved an estimated ten (10) test aircraft and the lives of four (4) test pilots in the past three years. Failure to receive these helmets will increase the possibility of this Center losing expensive aircraft and test aircrew personnel. FSC 8415-NL

(2) Research shows that the TOPTEX helmet has a marked superiority over similar helmets of competitive manufacturers. It will survive blows which would shatter other protective helmets and will distribute the applied force over a great area of the head, resulting in less brain injuries.

Signed: H. R. Plaga, Civilian RPO
UAL 952213D

5 From the personal archives of Mike Baetslé, Belgium

In a letter dated 20 May 1957, and signed by Dr. Charles F. Lombard[6], Director, Research & Development, Protection, Inc., we can examine the detailed product specification for the TOPTEX PILOT'S PROTECTIVE HELMET, MODEL T-3.

The scope of this specification covers requirements for the standard "TOPTEX" aircrew protective helmet manufactured by the Protection, Inc. Division of Mine Safety Appliance Company in accordance with principles set forth in U. S. Patent No. 2,625,683, a copy of which is shown on the right.

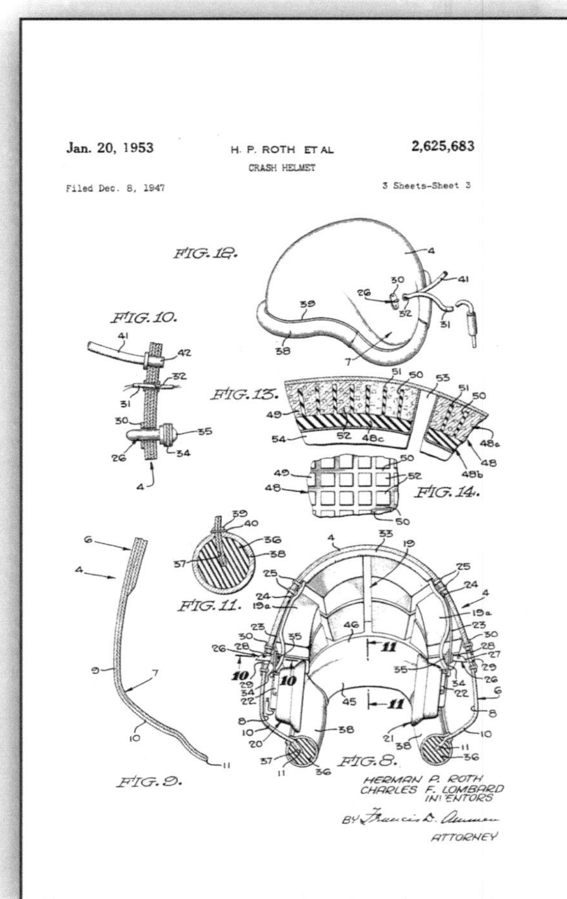

This helmet was intended to be worn in flight, and during escape, by pilots or crew members of aircraft, especially those whose maximum speed capabilities extend to Mach 1 or slightly above.

The helmet was produced in two basic types, custom fitted, and stock sizes. The custom fitted versions required a plaster replica of the wearer's head as a mold member in forming the interior cavity of the helmet liner, supplemented by specified head and face dimensions.

As of May, 1958, the TOPTEX T-3 was available in nine sizes of shells: SMALL sizes identified as 3AS and 3BS; MEDIUM sizes identified as AM, BM and SM; and LARGE sizes identified as AXL, AL, BL and NL.[7] This considerable sizing range extends far beyond the normal 95th percentile designed for by the Air Force personnel who have been unable to be fitted with standard Air Force flight helmets. However, examples of PI flight helmet tags from collectors reflect sizes of 3S and 3BM, which are not included in the original list of nine. This may indicate the possibility of updated sizing options.

[6] From the personal archives of Mike Baetslé, Belgium

[7] From the personal archives of Mike Baetslé, Belgium

It is also unknown why the helmet sizing was stamped on the model line of the PI tags.

The design and construction of the TOPTEX helmet shall be constructed of a high strength shell, molded of five ply fiberglass cloth laminated with plastic resin. Each shell shall be trimmed to basic dimensions for its type, with adjustment of trim dimensions to suit the face and neck dimensions furnished by the customer.

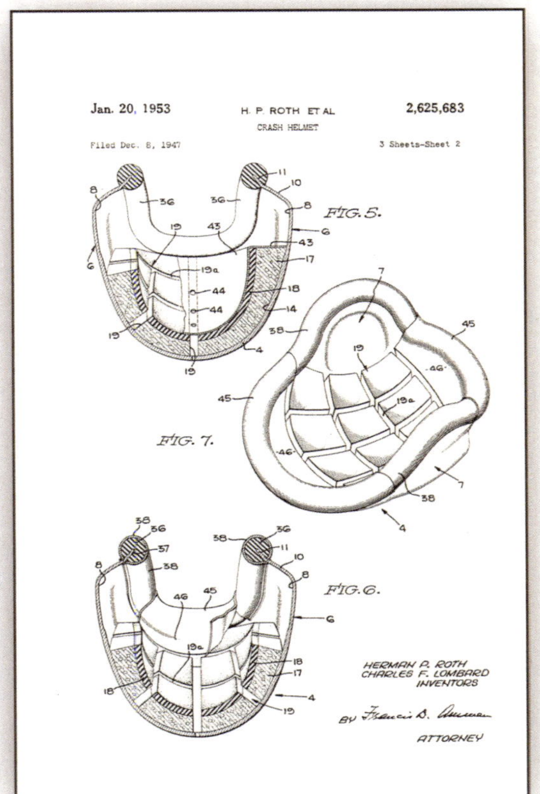

The standard finish was high quality white epoxy. A coat of Trewax or equivalent was specified as a final polish. The finished shell called for a protective edge roll of soft, open-cell latex foam rubber or equivalent, covered by soft, moisture-resistant leather.

The liner was fabricated in a mold corresponding to the shell size and type assigned to the customer, as determined by applicable head and face dimensions. The liner was comprised of two functional layers: 1.) A minimum thickness of one-half inch of substantially non-resilient cellular foamed plastic, having a compressive strength suitable for impact protection of the head; and 2.) An inner layer of soft, open-cell latex foam rubber molded to the plaster head form of the customer, or to a stock size form. The thickness of the foam rubber layer overlaying the non-resilient foamed plastic layer shall be between 1/8 inch and 1/2 inch. The liner shall be trimmed to integrate with the helmet shell and with the earphone cushions, and shall be provided with an interior surfacing of soft, moisture-resistant leather.

Without a doubt, these TOPTEX flight helmets far exceeded any standard issue helmets being produced at the time for the United States Air Force.Featuring a center-track, articulated, adjustable rigid wind-blast visor, this helmet set a standard of excellence that would not be matched in U. S. Air Force protective helmet design until the advent of the custom fitted HGU-26/P liners of the 1970s. Such is the reputation of the Protection Incorporated (later acquired by GENTEX) TOPTEX helmets that they remain especially sought after today by knowledgeable flight gear collectors and command a premium price, when they are encountered.

This original Thunderbirds helmet decal was used on the author's reproduction Thunderbirds TOPTEX flight helmet. *(Photo Credit: Author)*

Lets examine a number of TOPTEX flight helmets used by the Thunderbirds.

The Thunderbirds TOPTEX helmet is believed to have entered service in 1957 and used until 1964 (F-100C; F-105B).[8] This design change included six (6) symmetrical red "scallops" and maintained three (3) different sized dark blue stars. The classic Thunderbird emblem was also moved from the front to the rear of the helmet, where it remains to this day. The emblem detail was much improved, enlarged and added the now classic Thunderbirds "script". The emblem during this time period was a water slide decal, an example of which can be seen above. It is also noteworthy to point out that this decal includes an Air Force Outstanding Unit Award ribbon, the first of which was awarded to the team in 1958.

In this early example of the Thunderbird emblem, there is no gap between the red circled star, and the blue Thunderbird. In subsequent design changes, a slight gap or separation is incorporated.

[8] "Diamond in the Sky - A Pictorial History of the United States Air Force Thunderbirds"

United States Air Force Thunderbirds Flight Helmets

The first example of a TOPTEX-series Thunderbirds helmet I photographed was at the Southern Museum of Flight in Birmingham, AL (below) in the summer of 2019.

This helmet belonged to Captain Herman "GRIFF" Griffin, a solo pilot who flew from 1958 - 1961. This is the only example I have been able to locate that does not feature the 4-ship "Diamond"

formation inside the star on the Thunderbird emblem; rather a single F-100 for the solo pilot. This helmet also bears the Air Force Outstanding Unit Award ribbon, awarded in 1958. This would be the first of many awarded in the years that followed.

The star pattern is also interesting as through the next several decades, the time-tested pattern remained virtually unchanged, with the exception of a couple of deletions / additions flanking the unit citation ribbons.

The photos on this page are from a collector who wishes to remain anonymous, and is another example of a "GRIFF" helmet; however, this one includes the 4-ship "Diamond" formation inside of the star on the Thunderbirds emblem. This helmet is equipped with Hardman oxygen mask receivers. Unfortunately, there is no further information available on this beautiful example.

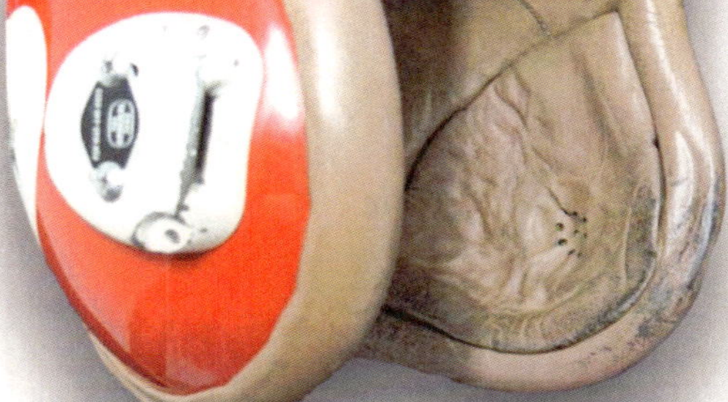

The example on this page comes from the same collector wishing to remain anonymous. The collector identified the pilot as Captain Gayle D. Williams, who flew the Left Wing position for the 1958 and 1958 show seasons. This helmet was equipped with later cast aluminum bayonet receivers, painted white.

The author is grateful for the use of these very rare helmet photographs from two pilots who were former teammates.

These photos on this page are courtesy of Lee Patterson of Arizona, who obtained this beautiful TOPTEX example from a team family member in the 1980s. Unfortunately, the pilot was not identified. The photo with the small star circled in yellow seems unique to this helmet and not consistent with any other TOPTEX helmets of the same period. These slight variations in star placement would be seen many more times in the years that followed.

Note: The visor arm attach point (yellow circle) "encroaches" on the white field of the helmet - something not seen on the previous examples.

United States Air Force Thunderbirds Flight Helmets

The stunning examples on this page are from the USAF Thunderbirds Museum at Nellis AFB, NV. The helmet belonged to Capt. Charles "Chuck" Maultsby, Right Wing, 1959 - 1960. Arguably, one of the best preserved and finest examples of this beautiful helmet, it is paired with an MBU-5/P oxygen mask.

Similar to those early P-1 and P-3 flight helmets, the TOPTEX helmets with the red scallops and blue stars on a white field closely mimic the paint scheme on the F-100 vertical and horizontal stabilizers.

One of the most significant improvements the TOPTEX helmet offered the wearer was a custom fit, leather-covered liner and edge roll — a big improvement over the P-series helmets. An open cell foam core liner was custom fitted to the wearer's head, then covered with soft leather, typically sheepskin. The earphones were also padded and covered in leather that provided the crew member with increased comfort as well as noise reduction.

Another improvement over the P-Series helmets was the elimination of the firm rubber edge bead that covered the exposed edge of the helmet shell. The edge bead was replaced with a soft foam edge roll covered with hand-stitched leather.

The result was aesthetically pleasing, but more importantly, provided the wearer with improved comfort. This style of edge roll and liner would be copied by flight helmet competitors and eventually become the industry standard — with only minor modifications — for the next several decades. This "imitation" was an example of the highest form of flattery!

TOPTEX flight helmet from Author's personal collection.

This TOPTEX helmet belonged to 1st Lieutenant William Pogue, Solo and Slot pilot, 1956 - 1957 show seasons. Pogue (retired Colonel) was also a NASA Astronaut on Skylab.

(Photo credit: Mike Moore)

The three photos on this page illustrate that despite all of these TOPTEX helmets being from the same period, many differences are noted.

As mentioned on the previous page, the scallops on the top photo appear (to my eye) to be much more "graceful" than the other two, which have a more linear appearance. In addition, the scallops on the middle example appear more "horizontal" than the top and bottom examples which tend to bear a more "downward" positioning.

The positioning of the scallops are also not consistent as you compare the layout based on the position of the bayonet receivers and the visor housing mounts. The scallops on all three examples are clearly different, proof positive that these helmets were without a doubt hand painted by individual artists.

Another inconsistency is the positions of the blue stars. The small stars flanking the "points" of the scallops are different in all three examples. Furthermore, The orientation of the large stars inside the scallops also vary, the bottom example pointing toward the arc of the scallop, the top and middle star pointing away from the arc.

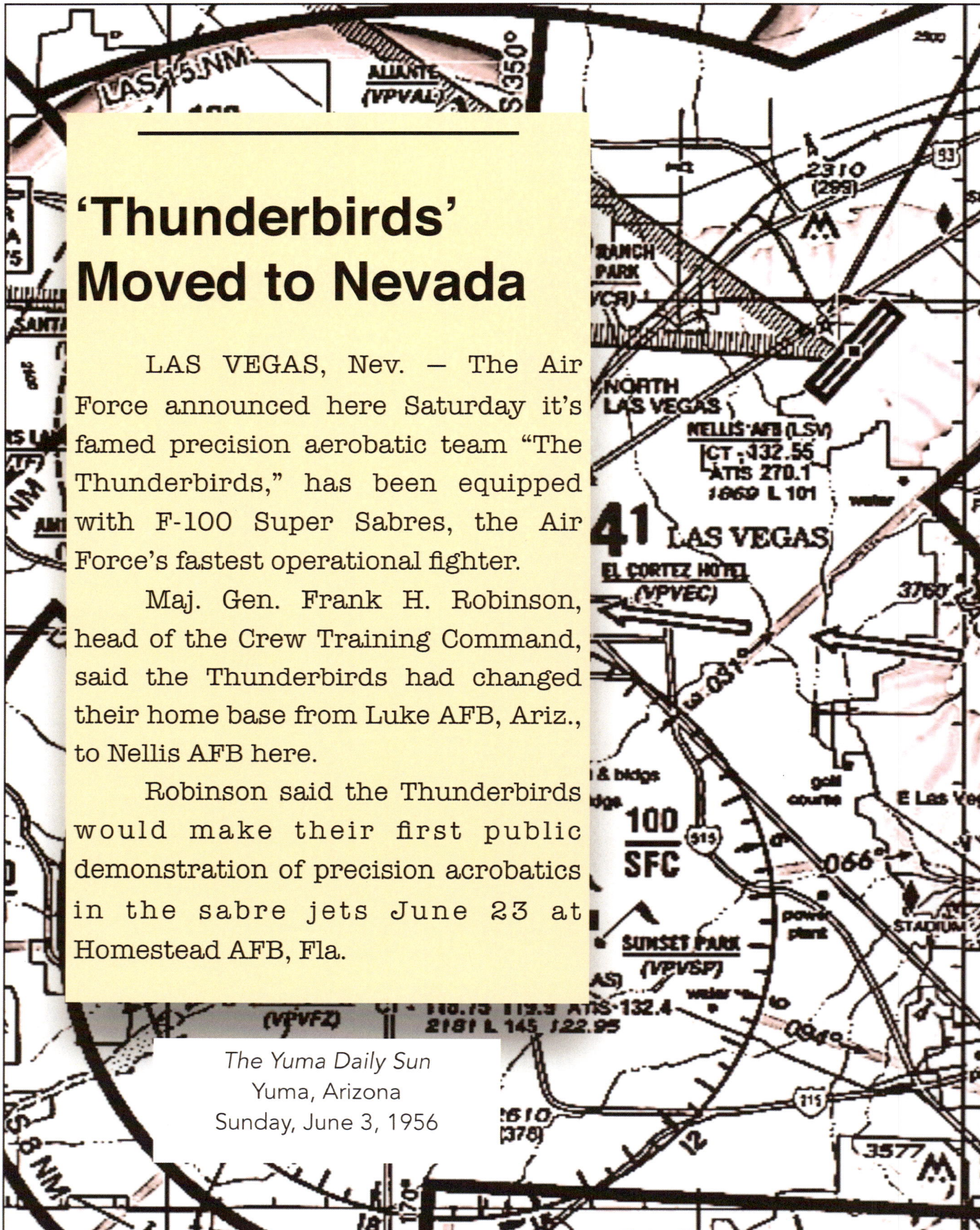

'Thunderbirds' Moved to Nevada

LAS VEGAS, Nev. – The Air Force announced here Saturday it's famed precision aerobatic team "The Thunderbirds," has been equipped with F-100 Super Sabres, the Air Force's fastest operational fighter.

Maj. Gen. Frank H. Robinson, head of the Crew Training Command, said the Thunderbirds had changed their home base from Luke AFB, Ariz., to Nellis AFB here.

Robinson said the Thunderbirds would make their first public demonstration of precision acrobatics in the sabre jets June 23 at Homestead AFB, Fla.

The Yuma Daily Sun
Yuma, Arizona
Sunday, June 3, 1956

"Members of the air force's famed 'Thunderbirds' precision demonstration who will perform Saturday at Geiger field[9] are (left to right) Capt. Samuel R. Johnson, solo pilot; Capt. Douglas D. Brenner, right wing; Maj. J. A. Robinson, leader; Capt. John M. Bartley, left wing, and 1st Lt. William R. Pogue, slot man. The aerial gymnastics, to start at 3:30 p. m., will be a feature of an open house celebrating the 50th anniversary of the air force[10]. The Spokane Air Force association and Ninth air division will sponsor."

---

[9] Now Spokane International Airport

[10] On 1 August 1907, an Aeronautical Division was established within the U. S. Army Office of the Chief Signal Officer (OCSO)

(NOTE: The Thunderbirds patches were also undergoing change, not just the aircraft!)

While still scarce, there are several examples of Thunderbird TOPTEX flight helmets in both museums and private collections. The TOPTEX helmets were likely in use between 1957 and 1964 — a total of 8 years. Multiply this by approximately 6 to 8 pilots per year,[11] and the numbers are still relatively low at less than 65 helmets.

Like their P-1 and P-3 predecessors, these helmets are highly collectable, not just by Thunderbirds enthusiasts and historians, but those knowledgable flight helmet collectors who know the rarity, and beauty, of *any* TOPTEX helmet.

[11] Educated guess based on available Thunderbird officer records, and the addition of a second solo pilot in 1962.

Undated photo of unidentified Thunderbirds pilots with their TOPTEX flight helmets.
(Photo courtesy of Lee Patterson)

Undated photo of unidentified Thunderbirds pilots with their TOPTEX flight helmets. (Photo courtesy of Lee Patterson)

Republic F-105B "Thunderchief"

Missing from this chapter are any references to the nearly forgotten F-105B "Thunderchief" that performed only six shows between April 26 and May 9, 1964. Following a tragic accident at Hamilton AFB, California that took the life of Capt. Eugene Joseph Devlin (second from right, below), the team transitioned back to the F-100 "Super Sabre" following the incident (from F-100C to F-100D).

This Team also wore TOPTEX flight helmets at the time, but it is unknown if the Thunderbird emblem on the back was updated to reflect the F-105 aircraft silhouettes. If so, these helmets would be as rare as the original P-1 and P-3 series flight helmets. Unfortunately, at the time of publication, 55 years had passed since the Team flew the "Thunderchief" and not a single photo of this rare helmet could be found.

USAF Thunderbirds, 1964 — F-105B
(left to right) Capt Clarence Langerud; Capt. Ron Catton; Capt Bill Higginbotham; Capt Jerry Shockley; Capt. Eugene Devlin; Maj Paul Kauttu.
(USAF Photo)

I reached out to Col. Michael "Mo" Beale, USAF (Retired) who is a former Thunderbird solo pilot (1989 - 1991), and former chairman of the Thunderbirds Alumni Association (TBAA). "Mo" contacted Col. Ron Catton, USAF (Retired) who flew in a solo position with that fateful F-105 Team in 1964. Col. Catton agreed to talk with me.

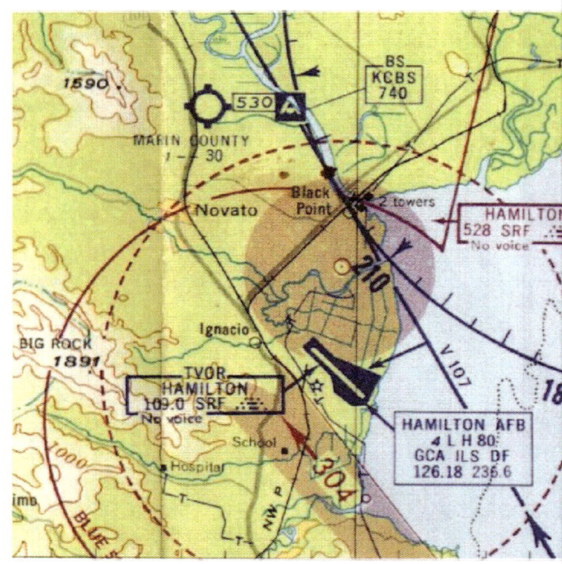

The date was 09 May 1964. The Thunderbirds had just completed their sixth air show in the Republic F-105 "Thunderchief" at McChord AFB, Washington. The Team flew to Hamilton AFB, CA, and at 6:41 PM, the formation was in the pattern. They descended towards the runway, and started a series of sequential pitch-ups by each aircraft to enter the landing pattern.

As Captain Eugene J. Devlin — the left wingman — pulled back on the controls, his aircraft broke up, and disintegrated. He did not survive.

I had the distinct privilege of speaking with Col. Ron Catton[12] USAF (Retired) in February 2019. At 86 years of age, he was sharp witted and

12 Captain Ron Catton, at time of service with the Thunderbirds

eager to share some phenomenal stories of his years as a Thunderbird. I was spellbound as I listened intently and respectfully!

I led off with prepared questions to confirm his time with the Thunderbirds, and the transition from the F-100C to the F-105B. I mentioned that the Thunderbirds were using the TOPTEX flight helmet while flying the F-100C and asked Col. Catton if they continued wearing them in the F-105B. He confirmed that they did in fact use the TOPTEX and that the custom leather liner and edge roll made the helmets far more comfortable than the standard USAF issue that most fighter pilots of that era were wearing.

When asked if he remembered the Thunderbird emblem on the back of the helmet, specifically asking if the aircraft silhouettes inside the star resembled the F-105B they were flying, his response was, "I honestly don't remember. I was a young Captain on top of the world at that time. Unfortunately, I didn't pay that close attention to this level of detail."

"I had a 1956 Corvette. I was a bachelor. And, my girlfriend was a dancer at the Tropicana."

Capt. Ron Catton
USAF Thunderbirds
1962 – 1964

Col. Catton shared that his association with the Thunderbirds predates his official time with the Team.

In the late 1950s, Col. Catton was stationed at Nellis AFB, and admired the Thunderbirds. He said, "Life was good. I had a 1956 Corvette. I was a bachelor. And, my girlfriend was a dancer at the Tropicana."

In 1961, he remembers receiving a phone call one day, asking him if he was available and willing to ferry Thunderbird #8 from Nellis AFB to Scott AFB, ahead of a scheduled Thunderbird performance. Of course, he jumped at the opportunity. He said it was pretty surreal strapping into that red, white and blue F-100C. The flight was uneventful until he got close to Scott AFB. The weather was bad. Remaining fuel was minimal.

Col. Catton contacted Air Traffic Control (ATC) as "Thunderbird #8" and announced he was landing at Scott AFB. Due to the deteriorating weather conditions, ATC asked for confirmation. Col. Catton affirmed his intentions, and a different controller came on the air, "Thunderbird #8...You're intention is to land at Scott AFB?" He replied, "That's affirmative." Col. Catton said the controller "Sounded like my father" and told him, "OK son, let's go to work!"

Since Col. Catton was flying in instrument meteorological conditions (IMC)[13] solely by reference to instruments, ATC used what is known as a ground control approach (GCA). The GCA controller uses radar to provide the pilot with course instructions and minor corrections (left and right) based on the extended centerline of the runway, as well as safe altitude "step downs" until reaching the runway threshold. The pilot is required to acknowledge each instruction until the "airport environment" is in sight, and a visual landing can be effected.

Col. Catton said he never saw the airport environment prior to ATC calling out one final instruction, "Thunderbird #8, raise your nose!" As soon as he reacted to the command, his main gear made contact with the runway! A success, but an example of what not to do! A bit rattled by the experience, Col. Catton secured the aircraft and made his way to the hotel for a well deserved night's rest.

---

[13] Also known as Instrument Flight Rules (IFR)

The following morning, Col. Catton got back into his perspiration stained flight suit and stopped in at the hotel coffee shop. Much to his surprise, the Thunderbirds were also in the shop, impeccably dressed in their pristine show suits and silk ascots. He didn't feel properly dressed for such an occasion!

The Thunderbird commander, Major Ralph "Hoot" Gibson[1], called out, "Ron, what are you doing here?" "I brought in an aircraft for you last night, Sir" was the response. Col. Catton said he was hit with a barrage of follow up questions, "Was it raining?" "Was it below minimums when you landed?" As a concerned commander, Maj. Gibson knew the answers, but eventually backed off.

Col. Catton stayed at Scott AFB and watched the Thunderbirds' performance. Following the show, Maj. Gibson asked Col. Catton if he could fly one of the Thunderbird aircraft back to Nellis AFB with the rest of the team. Again, how could one turn down such a rare invitation? The answer was a resounding "Yes, Sir!"

The flight home required a fuel stop in Albuquerque, NM. Once again, the approach was plagued by heavy thunderstorms, increasing the stress and workload to get the aircraft safely on the ground. ATC had the Thunderbirds staged in a holding pattern, which is a racetrack shaped course over a specific point, with aircraft stacked at various altitudes, while waiting for favorable landing conditions. Slowly, the Thunderbirds were assigned lower altitudes, and in two-ship packages, were cleared for approach and landing.

[1] Not the NASA Astronaut.

As Col. Catton was cleared to land, his lead experienced a complete radio failure, requiring Catton to take the lead. Making matters worse, Maj. Kattu reported 1 to 2 inches of standing water on the runway.

When landing in formation, Col. Catton explains that after both aircraft touch down, the trailing aircraft deploys its drag chute first, then calls out to the lead, "Good chute" which is the signal for the lead to deploy his drag chute. This is a safety procedure. Hypothetically, if the lead deployed their chute first, and the trailing aircraft experienced a chute failure, the trailing aircraft could possibly overtake or worse, run into the lead aircraft.

Col. Catton lands and instinctively waits for his wingman to call out, "Good chute." Radio silence. Then he remembers, his wingman has no radio!

Despite the safety protocol, Col. Catton decides its time to deploy his chute before he runs out of usable runway. When he pulled the handle, the entire drag chute and the canister it is housed in came off the aircraft, dropping it onto the wet runway! He cautiously applied the brakes, but they "grabbed" causing the wheels to hydroplane, and spinning the aircraft 180 degrees. As his F-100C continued hurtling down the runway — backward — Col. Catton had an epiphany: "Add power!" His quick thinking worked, and he regained control of the aircraft. Another weather disaster averted.

When Col. Catton and the rest of the Thunderbirds arrived back at Nellis AFB, Maj. Gibson called to offer his appreciation for Catton's assistance. Maj. Gibson asked, "Have you ever thought about being a Thunderbird?" Col. Catton said all USAF pilots have thought about being a Thunderbird at one time or another, but he told Maj. Gibson he didn't think his fitness report was up to Thunderbird standards. Col. Catton was flattered that Maj. Gibson told him, "I know you better than anyone" and insisted he apply.

Also applying for the team at the time was Maj. Paul Kauttu[15] — a good friend who would become Catton's Best Man at his wedding. Col. Catton said he remembers receiving a call from the Thunderbirds informing him he had been selected to the Team as a solo pilot. Maj. Kauttu was also selected for the slot position, but the Thunderbirds were unsuccessful in contacting him with the announcement. Col. Catton said he meets Kauttu every morning and would be happy to deliver the message. The following morning, over breakfast, Catton revealed the news. Overcome with excitement, the two literally danced in the restaurant!

The stories continued. Col. Catton asked if I knew the name Bob Hoover. "Yes, I do." For those who may not be familiar, Bob Hoover is known among aviation circles as a "pilot's pilot" and by others, as the greatest pilot to ever have lived. Whatever your personal opinion, one can't deny Bob Hoover's many accomplishments: United States Army Air Forces fighter pilot, USAF and civilian test pilot, flight instructor, air show pilot, and aviation record-setter. If that wasn't enough, Hoover was also Gen. Chuck Yeager's backup pilot for the sound barrier breaking X-1 flight.

Bob Hoover was also an F-100 test pilot as well as a spokesperson for Republic Aircraft and would attend air shows to promote the aircraft and the company.

At a show in Reno, NV in 1962, Bob Hoover performed in a P-51 "Mustang" followed by the Thunderbirds. At the end of the day, the pilots were relaxing, "rehydrating" and discussing performance challenges in the F-100. Col. Catton said he, Hoover, and Thunderbird Right Wing pilot, Capt. Ralph Brooks were comparing notes. Catton described Brooks as the best Thunderbird pilot in their era, and one of the best pilots he ever knew.

The three pilots were discussing the difficulty of recovering from a high-speed dive in an F-100. Hoover asked Capt. Brooks, "You're in a high-speed dive…you know you're going to die…what would you do?" Capt. Brooks responded, "Full afterburner…Stick

---

15 After flying the slot position for one year, Maj. Kauttu became the Leader for the subsequent two seasons.

full aft...nose above the horizon, and hang on!" Hoover asked Catton the same question, who agreed in principle with Brooks' response.

Hoover doesn't disagree with their assessment, but adds, "Just before hitting the ground, I would jam the stick forward!" When I heard this, I said, "What? That's counterintuitive!" Col. Catton said, "That was my exact response!"

Hoover surmised that under those conditions, the aircraft would theoretically "Bounce off of ground effect![16]" The debate over theory and practice have always been real. But in my opinion, this sounded crazy!

> **"Just before hitting the ground, I would jam the stick forward!"**
>
> Bob Hoover

A couple of weeks later, the Thunderbirds were performing an air show in Pampa, TX. The solos were performing an opposing loop, and came over the top and Col. Catton found himself headed straight for the ground at a very high rate of speed. He said his first thought was, "I'm going to die!" He was in full afterburner, stick against his gut, nose above the horizon, still descending! Bob Hoover briefly came to mind. Catton said, "I closed my eyes...briefly hesitated...then jammed the stick full forward with both hands!" When he opened his eyes, he was so low over a Texas corn field, he was afraid to pull up for fear of hitting the tail on the ground! Sounds like Hoover got it right!

Pampa, TX was a small airport, so the Thunderbirds flew in and out of Amarillo, AFB, TX.[17] When they completed the show and got back on the ground, his crew chief threw up his hands, flabbergasted at what he was witnessing. When Col. Catton deplaned, he noticed the bottom of his F-100 was stained bright green and there were remnants of corn stalks stuck to the air intake!

---

[16] The effect of added aerodynamic buoyancy produced by a cushion of air below an aircraft moving close to the ground.

[17] Closed in 1968.

Col. Catton hastily made his way to the hotel to soothe his frazzled nerves with a bottle of Jack Daniels. After some self-medicating, Lt. Col. William Alden (Thunderbirds Commander) stormed into Catton's room shouting, "Catton, you almost bought it!" Catton uttered, "Yes, Sir!" Alden followed up with, "You want to quit?" "No, Sir" replied Catton. Alden asked Catton if he had any questions. "Just one, Sir. What did it look like?" Alden said, "Like a rabbit going through a hay field!"

Col. Catton pressed on and got right back in the saddle. But not without cheating death one more time.

The Thunderbirds were performing an air show at Randolph AFB, TX (San Antonio) and Catton was performing a slow speed half Cuban 8 — a maneuver Catton developed specifically for the F-105B. As he comes over the top, the aircraft goes into an unintentional inverted spiral. The control inputs were unresponsive. At about 1,500 feet above ground level, the aircraft completes a 270 degree turn in full afterburner, recovering with minimal ground clearance. This recovery, coupled with the afterburner necessary to rejoin the delta, created a critical fuel situation and Catton had to land, cutting his portion of the show short.

"Lead...Catt's on fire!"

Capt. William Higginbotham
USAF Thunderbirds
1963 – 1964

Of course, this story doesn't end here. Upon landing, his drag chute failed. Again. He applied heavy braking and the anti-skid brakes failed. The main tires blew. Just when he thought the ordeal was over, Capt. William Higginbotham, Right Wing called out over the radio, "Lead...Catt's on fire!" It turns out that the brakes overheated and the magnesium wheels caught on fire. Col. Catton said he opened the canopy and vaguely remembers jumping out of the cockpit! He walked away from another averted disaster.

F-105B solos performing the "Calypso Pass."
(Photo credit: USAF)

Following the show, everyone was gathered in the Officer's Club at Amarillo AFB for customary "rehydrating" and camaraderie. Bob Hoover approached Col. Catton looking "angry" but as he advanced closer, began to smile. Hoover reportedly said, "Catton, the least you could have done was bail out!" Hoover continued, "I thought you were going to die! The crowd thought it was a planned maneuver!"

Before we concluded our conversation, Col. Catton agreed to recount his memory of the accident that took the life of Captain Eugene J. Devlin.

The Team had just completed an air show at McChord AFB, only their 6th in the F-105B. Capt. Devlin approached Maj. Paul Kauttu, Leader to report that his aircraft was making what he described as "strange noises" during their last performance. There were no other indicators pointing to the likely cause. Maj. Kauttu decided the

Team would fly to Hamilton AFB, forgo the arrival maneuvers to minimize any unnecessary stress on the airframe, land, and have maintenance perform a complete inspection of his aircraft.

Upon arrival at Hamilton AFB, Maj. Kauttu pitched up for landing, followed by Capt. Devlin, who was flying the #2 Left Wing position. Col. Catton said he and Capt. Clarence Langerud, opposing solo, were trailing the Thunderbird diamond formation. Col. Catton was inverted, mirroring Capt. Langerud's position, a maneuver known as the "Calypso."

As Capt. Devlin pitched up for landing, his aircraft disintegrated. Catton and Langerud were following 200 to 300 yards behind and flew through the debris field before pitching away. Col. Catton thought he saw Capt. Devlin's parachute, but it turned out the aircraft drag chute deployed upon breakup. Capt. Devlin did not survive.

Later that evening while in the Bachelor's Officer's Quarters (BOQ), there was a knock on the door. It was a civilian holding a paper bag. He said, "I have film of your arrival. I would like you to have the film…it might be useful in the accident investigation."

According to Col. Catton, the accident investigation revealed the trapezoidal stress plate, located behind the cockpit on Capt. Devlin's aircraft suffered a catastrophic failure, resulting in the inflight destruction of the aircraft. When the remainder of the Team returned to Nellis AFB, they turned their aircraft over to Republic for detailed inspections; however, the Thunderbirds would not continue their season in the F-105B.

Following a stand down period of a couple of months, the Team went back to the F-100, but not the "C" model. Instead, they chose the upgraded "D" model, which had flaps — a big improvement in increasing lift and reducing speed during landing. Col. Catton concluded his service with the Thunderbirds at the conclusion of the 1964 show season.

In closing, I asked Col. Catton if he had any photos to share of his Thunderbird flight helmet. His response was to the point, "No." So, I asked, what ever happened to that

TOPTEX flight helmet with what I want to believe are silhouettes of F-105s on that beautiful Thunderbird emblem. Honestly, I was not prepared for his response. Col. Catton said he liked the helmet so much, that when he separated from the Thunderbirds, he had his helmet repainted back to stock white so he could use it in his subsequent flying assignments!

I have no words!

Capt. Ron Catton performing the "Dead Ant" at altitude.
(Photo credit: USAF)

As a side story, Col. Catton mentioned that he developed a maneuver specifically for the F-105 called the "Dead Ant."

The maneuver involved an inverted, low level pass with the landing gear extended; however, it was never included in their regular air show routine due to safety concerns.

With every story comes a story. Col. Catton said he introduced this maneuver to mimic a Canadian "drinking game." When someone in the bar shouts, "Dead Ant," Everyone is expected to drop to the floor on their backs, arms and legs extended towards the ceiling, looking like dead ants. The last one in position is required to buy a round for the house!

What may have been. The author's artist conception of the USAF Thunderbirds TOPTEX flight helmet decal, F-105B.

PROTECTION, INC. DIVISION

PRODUCT SPECIFICATION

SP-10

TOPTEX PILOTS' PROTECTIVE HELMET, MODEL T-3

May 20, 1957

Approved Charles F. Lombard
Charles F. Lombard
Director, Research & Development

REVISIONS

Appr.

No. 1 - Nov. 15, 1957; General revision
No. 2 - Nov. 22, 1957; p. 3: change in (1) resin content and (2) limits of foam rubber thickness; p.4: sec. 3.1 modified; p.5: sections 3.2 and 3.3 modified.

PROTECTION, INC. DIVISION

SPECIFICATION SP-10

TOPTEX PILOTS' PROTECTIVE HELMET

Model T-3

1. SCOPE

1.1 This specification covers requirements for the Standard "Toptex" aircrew protective helmet manufactured by the Protection, Inc. Division of Mine Safety Appliances Company in accordance with principles set forth in U. S. Patent No. 2,625,683.

1.2 Intended Use - The type of helmet covered by this specification is intended to be worn in flight, and during emergency escape, by pilots or crew members of aircraft, especially those whose maximum speed capabilities extend to Mach 1 or slightly above. Realizing that design of articles of this kind must represent compromise between conflicting factors, this helmet is intended to offer an optimum balance of design factors contributing to:

(1) Functional efficiency during normal flight conditions; and

(2) Protection from air blast, impact and other hazards during emergencies.

1.3 Types and Sizes

1.3.1 This helmet shall be available in two types: (a) Custom-fitted and (b) Stock sizes. Basic design and construction shall be the same for both types, but only the custom-fitted type is recommended where the following factors are critical:

(1) Fit, stability and comfort;

(2) Maximum protection and retention during escape at high speed.

1.3.1.1 Custom-fitted helmets shall be fabricated by a method employing a plaster replica of the head as a mold member in forming the interior cavity of the helmet liner, supplemented by specified head and face dimensions.

SP 10 Revision No. 2 - 1 - 11/22/57

2. DESIGN AND CONSTRUCTION

2.1 General - The helmet shall consist of the following basic components:

(1) A high strength shell provided with leather-covered cushioned trim on exposed edges for maximum helmet stability and noise exclusion, and with an exterior finish having maximum practical heat reflection.

(2) An inner liner comprising (a) an impact energy-absorbing layer of crushable cellular plastic; (b) a sizing and comfort layer of soft, resilient, foam material, with soft leather covering on the interior surface.

(3) Communications equipment (when ordered) consisting of (a) earphones mounted in leather-covered, raised-edge cushions; (b) electrical cordage, with plug, leading to earphones and (where specified) an externally-mounted connector for mask microphone.

2.1.1 Accessories - The following accessory equipment shall be available:

(1) A protective visor for wind blast and for glare protection, which may conveniently be secured in either raised or lowered position; and

(2) Alternative provisions for oxygen mask attachment, consisting of either: (a) leather tabs with snaps for use of past conventional oxygen mask harnesses (recommended for use with aircraft with speed capabilities not exceeding the 200-300 knot range); or (b) the Toptex* oxygen mask retainer kit.

2.2 Materials and Workmanship - Materials shall be as specified on applicable drawings or as specified herein. When materials are used which are not specifically designated, they shall be entirely suitable for the purpose intended, and of the lightest practicable weight. Workmanship shall be in accordance with high grade practice covering this type of work.

*Supplied to the Armed Services as the "Hardman" retainer.

2.3 Detail Construction

2.3.1 Shell - Details of shell construction shall be the same regardless of shell size except for necessary minor dimensional differences. A suitable range of shell sizes shall be available to accommodate the complete range of normal adult human head sizes.

2.3.1.1 Molding - The shell shall be molded of fiberglass cloth laminated with plastic resin in accordance with high quality commercial practice. The shell structure shall comprise a minimum of five plies in any area. There shall be no butt joints in plies, and typical overlap shall be one inch, with a minimum at any point of one-half inch. The polyester resin used for lamination shall be Interchemical No. 625, or equal, used with suitable catalyst and accelerator. The laminate shall be well saturated with resin, substantially void-free, and shall have a resin content, by weight, between 35 and 45 per cent, following cure using a vacuum bag technique.

2.3.1.2 Trimming and Drilling - Each shell shall be trimmed to basic dimensions for its type, with adjustment of trim dimensions to suit face and neck dimensions furnished by the customer as closely as possible. Drilling shall be performed using properly sharpened tools, with support where possible, to assure clean holes and avoid local delamination. All edges from trimming or drilling shall be smoothed.

2.3.1.3 Finishing - The standard finish for shells shall be a high quality white epoxy, Magna Coatings 3-W-8 Type III, applied over 3-W-24 Surfacer, or equivalent. Manufacturer's instructions shall be followed to insure maximum adhesion and a smooth finish. A coat of Trewax or equivalent shall be used as a final polish.

2.3.1.4 Edge Roll - The finished shell shall be provided with a protective edge roll of soft open-cell latex foam rubber or equal, covered by soft, moisture-resistant leather.

2.3.2 Liner - Each liner shall be fabricated in a mold corresponding to the shell size and type assigned to the customer, as determined by applicable head and face dimensions. The liner shall comprise two functional layers: (a) A minimum thickness of one-half inch of substantially non-resilient cellular foamed plastic, having a compressive strength suitable for impact protection of the head; and (b) an inner layer of soft, open-cell latex foam rubber molded to the plaster head form of the customer or to a stock size form. The thickness of the foam rubber layer overlying the non-resilient foamed plastic layer shall be between one-eighth and one-half inch. The liner shall be trimmed to integrate with the helmet shell and with the earphone cushions, and shall be provided with an interior surfacing of soft, moisture-resistant leather.

PROTECTION, INC. DIVISION

2.3.3 Phone Cushions - The helmet shall be provided with a pair of soft, molded latex foam rubber phone cushions having raised edges cooperating with the shell edge pad to contact the face and neck as a barrier to external noise. The cushions shall provide means for mounting phone units of various types as required by aircraft communications systems. The cushions shall have a surface of soft, moisture-resistant leather.

2.3.4 Communications Equipment - The helmet shall be provided with communications equipment and cordage, as specified and/or supplied by the customer, appropriate to the communications system of the aircraft with which the helmet is to be used. The cordage shall enter the helmet through a protective grommet located as required by the type of cordage used, or otherwise as specified by the customer. When the cordage includes a jack for connection of mask microphone, this shall be securely mounted on the exterior surface of the shell in a convenient location which may be customer-specified when desired, and retained by a molded protective cover stitched to the shell through drilled holes.

3. PERFORMANCE REQUIREMENTS

3.1 Impact Resistance - Impact resistance of the helmet shall be such that when tested by Protection, Inc. Division impact test equipment and procedures specified in paragraph 3.1.1, peak resultant acceleration of the test head shall not exceed 120g, with the rate of change of g not exceeding 150 kilo-g per second. (Note: This is not a limiting tolerance figure, as physiological tolerances to impact have not been satisfactorily established.)

3.1.1 Standard helmet test equipment and procedure shall be as follows:

(1) Impacting mass: 1/2 slug (16.08 lb.).

(2) Striking surface of impacting mass: 1.9 in. spherical radius.

(3) Velocity for qualifying impact: 16 ft. per second.

(4) Impact site (for qualifying impact): side of helmet, approximately 45 degrees off vertical axis. A previously untested specimen shall be used, of a size providing satisfactory fit on test head.

(5) Test head: 12.25 lb. cored magnesium casting with c.g. approximating that of human head, provided with internally machined mounting areas for attachment of an accelerometer directly beneath chosen impact site. Head unrestrained during and after impact.

(6) High response sensing and recording instrumentation

SP 10 Revision No. 2 - 4 - 11/22/57

2800 3/57 T-3142

capable of true sensing and recording transient impact phenomena typical of total impact periods of the order of .004 seconds and peak "g" values to 400 or 500 "g".

3.2 Piercing Resistance - Piercing resistance of the helmet shall be in excess of that required by current issues of Specifications MIL-H-9480 (USAF) or MIL-H-1966 (Aer).

3.3 Wind Blast Protection - A typical sample of the helmet equipped with the Toptex mask retainer kit and A-13A mask, and the visor lens trimmed to integrate properly with the mask and retainer shell, shall be properly mounted on and fitted to an anthropometric dummy. The dummy shall be secured to an aircraft ejection seat mounted in a test cockpit on a rocket sled, and ejected at an air speed between Mach 1 and Mach 1.1. The visor shall remain down and locked, and the helmet shall remain on and protect the head during the exposure to wind blast during such ejection.

3.4 Acoustic Characteristics - The helmet shall provide maximum possible exclusion of external noise, and maximum effective transmission of intelligible signals from the earphone to the ear of the wearer, consistent with the existing state of the art.

(NOTE: At the time of preparation of this specification, there are no military or other generally accepted standards for acoustical testing and performance of helmets. Whenever practical, such standards will be incorporated in this specification.)

3.5 Thermal Characteristics - The helmet shall provide maximum practicable reflective rejection of external radiant heat from the surface of the helmet shell, in combination with a liner having good insulating qualities against conducted heat.

(NOTE: At the time of preparation of this specification, there are no military or other generally accepted standards for thermal testing and performance of helmets. Whenever practical, such standards will be incorporated in this specification.)

The Protection, Inc., TOPTEX Model T-3 document on the preceding 6 pages is courtesy of, and from the personal archives of Mike Baetslé, Belgium.

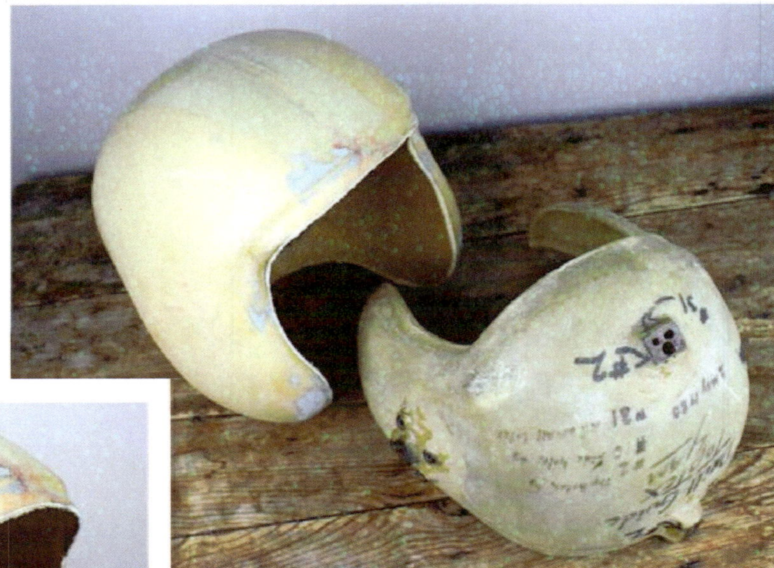

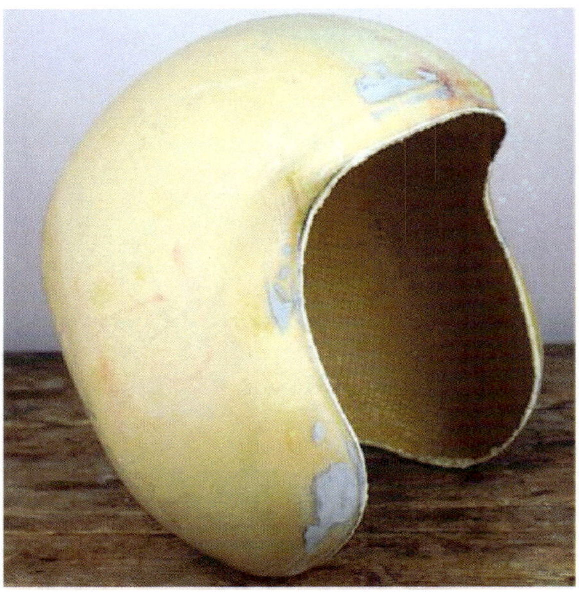

These photos of a TOPTEX helmet drill guide are courtesy of, and from the personal archives of Mike Baetslé, Belgium.

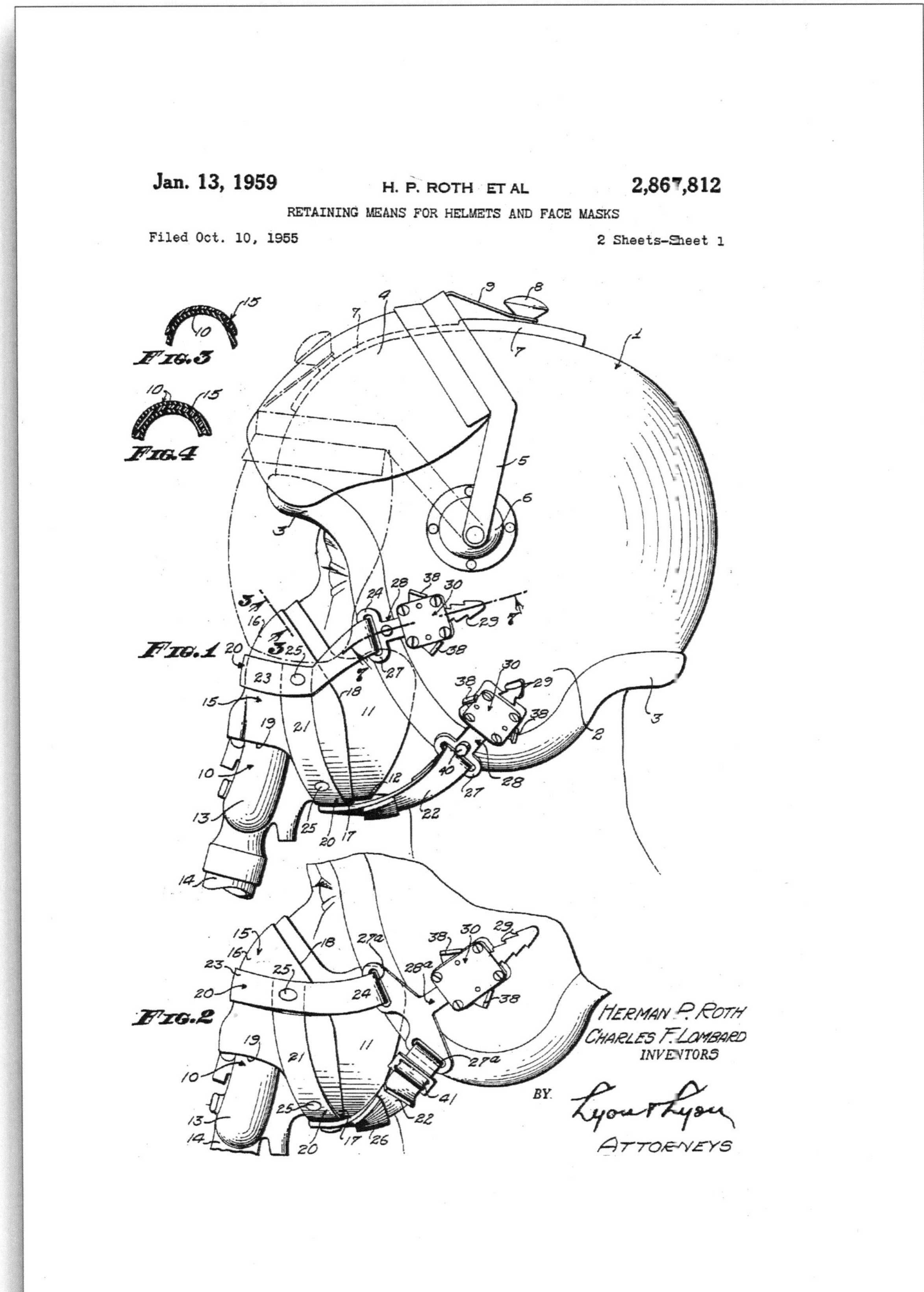

Jan. 13, 1959 H. P. ROTH ET AL 2,867,812

RETAINING MEANS FOR HELMETS AND FACE MASKS

Filed Oct. 10, 1955 2 Sheets-Sheet 1

HERMAN P. ROTH
CHARLES F. LOMBARD
INVENTORS

BY Lyon & Lyon
ATTORNEYS

5

HGU-2A/P & HGU-26/P Series Flight Helmets

The early HGU-series[1] of flight helmets were first introduced in 1959 and saw widespread use throughout the USAF in the 1960s and 1970s. First produced by Consolidated Controls, contracts were subsequently awarded to GENTEX, Sierra Engineering Corporation and Land Manufacturing Company.

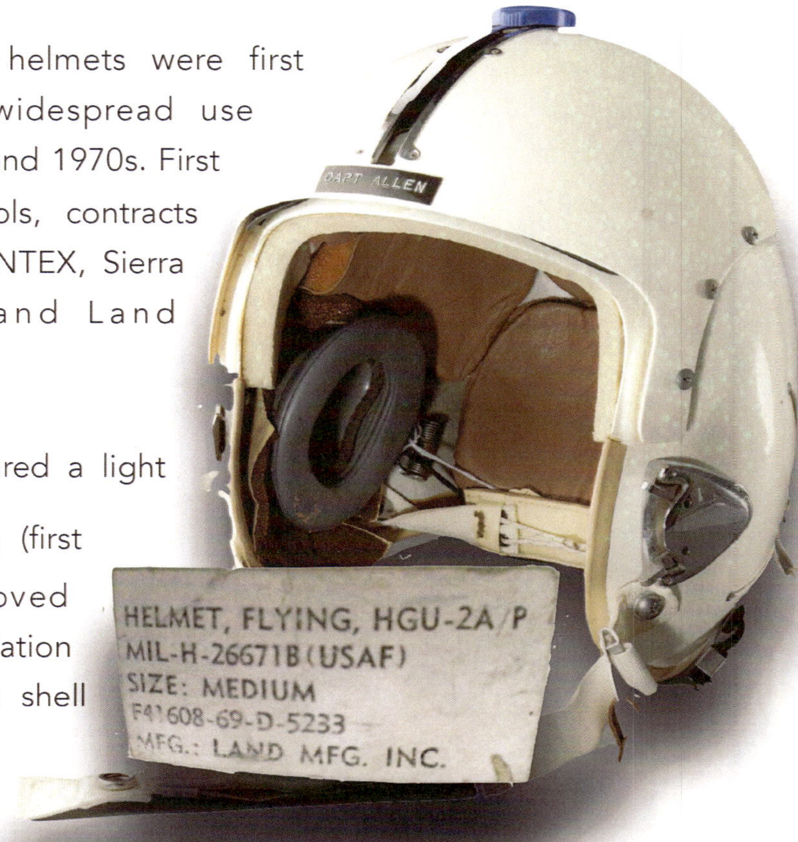

Helmet, Flying, Protective, Type HGU-2A/P, USAF
(Photo credit: Smithsonian National Air and Space Museum)

The HGU-2/P and HGU-2A/P featured a light fiberglass shell[2], fixed visor housing (first fiberglass, then plastic), improved headsets and typically three ventilation holes drilled in the crown of the shell under the visor housing to allow heat from the wearer's head to escape. Adjustable fitting pads were also introduced to customize and improve the fit and comfort of the basic helmet.

By the mid-1970s, the HGU-2/P and HGU-2A/P nomenclature became obsolete and were re-designated the HGU-26/P for both single and dual visor options (photo on right).

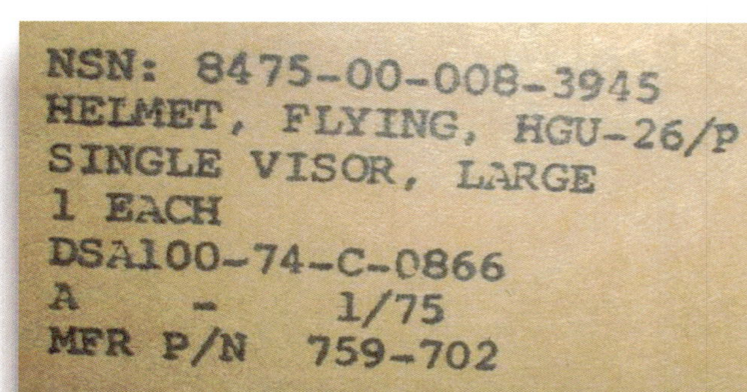

[1] HGU: Helmet General Usage

[2] The basic helmet shell, referred to as a Quarter Assembly, is designated HGU-22/P

As the USAF transitioned to the HGU-series flight helmets, so did the Thunderbirds. These helmets were in service between 1964 and 1982 and used in the F-100D "Super Saber" (later years), F-4E "Phantom" and T-38A "Talon."[3]

Beginning with the TOPTEX helmets, the Thunderbirds emblem on the rear always captured the silhouettes of the aircraft the team was currently flying. This practice continued with the HGU-series helmets, unfortunately, there are no clear reference photos showing the F-100s on the HGU-style helmet, only the F-4s and T-38s shown below.

F-4E "Phantom"

T-38A "Talon"

As seen in these photos, a couple of slight design changes occurred with the transition between the F-4 and the T-38 including a white "space" between the blue Thunderbird and the red star / circle, a lighter shade of blue, and the "T" in the Thunderbirds script; however, the latter was not consistent on all T-38-era helmets.

On the next page are a couple of photos of F-100D "Super Saber" Thunderbird pilots with HGU-style flight helmets. The red scallop design, star pattern and Thunderbird emblem placement appear unchanged from the TOPTEX helmets.

[3] Diamond in the Sky - A Pictorial History of the United States Air Force Thunderbirds

The example on the right (USAF Photo) features Capt. Russell Goodman, Narrator for the 1964-1965 show seasons.[4] His helmet features a blue visor lock knob, white (painted) cast aluminum oxygen mask receivers, and a white chin strap, the latter of which would

remain unchanged for many years. Another interesting note are the large stars inside the red scallops are pointed away from the inside of the radius, opposite of the example in the black and white photo. Unfortunately, we can't see the Thunderbird emblem and the F-100 "Diamond" formation. The star "orientation" was not consistent in the years that followed.

The black and white photos (USAF Photos) are of Maj. Neil Eddins, Commander and Leader for the 1967-1968 show seasons, and provides us with a rare glimpse of the back of his helmet. In this example, the large stars inside the scallops are positioned with the "points" towards the inside of the radius; however, one again, the F-100 silhouettes are not clear.

[4] Photo credit: AerobaticTeams.net.

The beautiful HGU-26/P helmet with MBU-5/P oxygen mask (below) is on display at the Castle Air Museum, Atwater, CA. Another small change can be noted on the visor lock knob — early models were blue before switching to white. This helmet, and the one on the previous page, also feature white, aluminum cast oxygen mask receivers installed to accept "T" style bayonets.[5]

The F-4-era HGU-26/P flight helmet on the next page is on display at the Smithsonian National Museum of African American History and Culture. It was worn by, and is a gift

from Gen. Lloyd "Fig" Newton, who served with the USAF Thunderbirds as the Narrator as well as the Slot and Right Wing positions between 1975 and 1978. Notice the large stars are "pointed" toward the arc of the scallop. Also note the call sign "Fig" painted below the visor track, common to this period.

5 The oxygen mask is attached to the "T" bayonet with webbing, and secured to the helmet via the oxygen mask receiver.

I was very curious as to why Gen. Newton would have a flight helmet bearing the silhouettes of F-4s when he first joined the team in 1975, after the Thunderbirds transitioned to T-38s. I reached out to Brig. Gen. Dan Cherry, Thunderbird Commander / Leader for the 1977 and 1978 show seasons — and good friends with Gen. Newton — looking for answers.

According to Gen. Newton, he was selected for and joined the Team in late 1974 to serve as narrator in 1975. At that time, the Air Force was in the process of transitioning from the F-4 to the T-38. When Gen. Newton joined the team in 1974 they were still flying the F-4 so the helmet he received logically had the F-4 silhouettes.

Another example of an F-4-era helmet is shown on the left, minus the unit citation ribbon. This was a presentation helmet to SSgt. Ray Herder, who served as a USAF Thunderbirds Crew Chief between 13 February 1973 and 15 December 1975.[6]

Note "Ray" painted below the visor track.

The helmet on this page is an example of an F-4-era flight helmet made by Protection, Incorporated.[7] This example is also void of any unit citation

[6] Photos courtesy of John Slezak.

[7] Protection, Incorporated was purchased by General Textile Mills (GENTEX) in 1969.

ribbons and, curiously, the dot above the "i" in Thunderbirds is missing. Also note the use of blue and white nylon chin straps on the examples that follow.

These helmets also reflect another change from white painted, cast aluminum oxygen mask receivers and "T" bayonets, to flashier, chrome plated bayonet receivers with updated "J" style bayonets, which provided a better seal of the oxygen mask against the wearer's face. The photos on this page detail the chrome oxygen mask receivers and the "J" bayonets.

The USAF issued HGU-2 and HGU-26 series flight helmets were very basic with plain rubber edge beading (as shown on the first page) and a styrene liner with a series of leather faced foam fitting pads attached with velcro at the forehead, nape and crown, shown on the left.

Detail showing the "J" Oxygen Mask Bayonet and the Chrome Plated, cast aluminum Oxygen Mask Receivers. (Photo credits: salembeti.com)

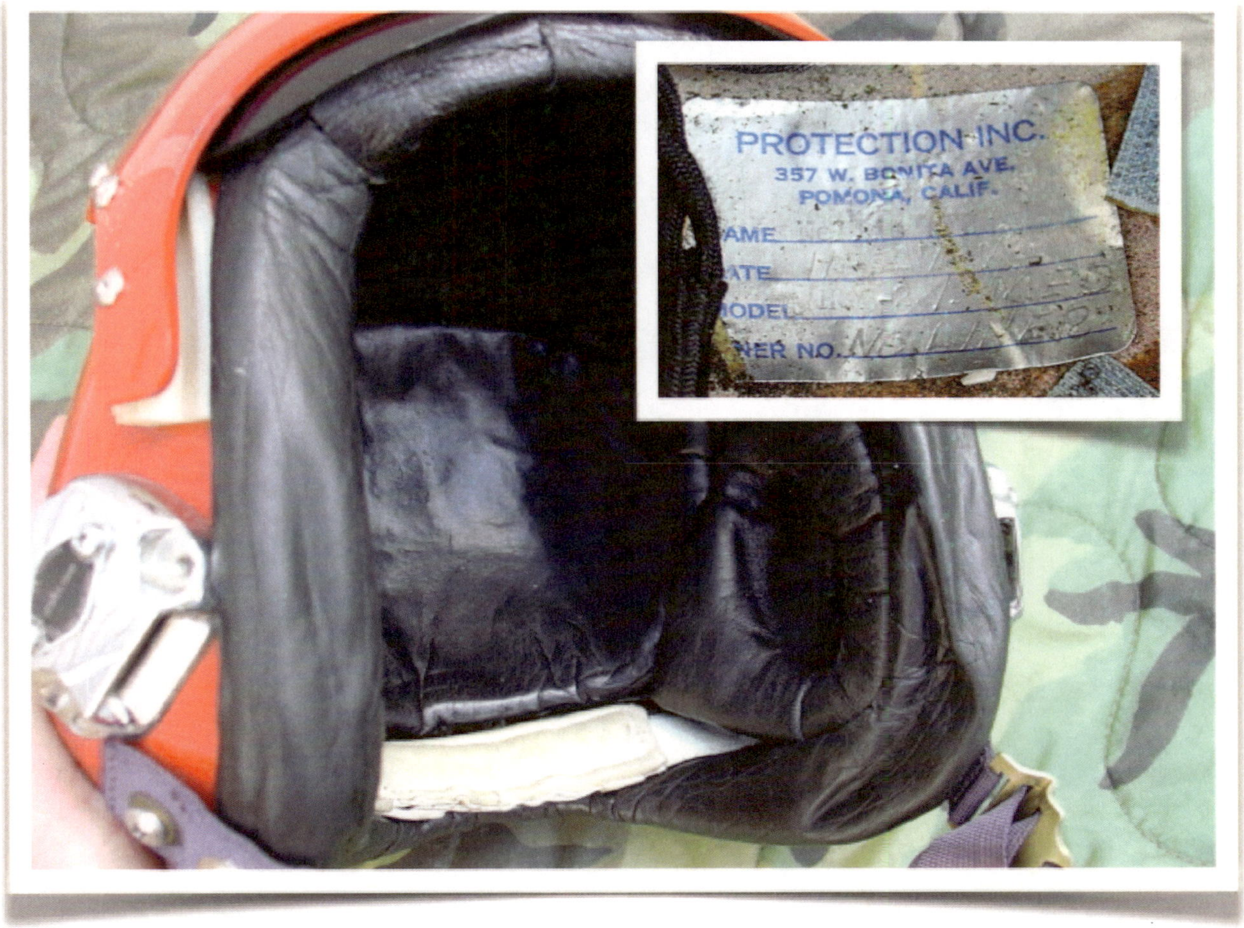

In contrast, the helmets worn by the Thunderbirds were built by Protection, Inc. during this period and afforded the wearer many custom upgrades few enjoyed in the "regular" Air Force. Thunderbirds pilots were treated to hand sewn, leather covered, foam edge rolls, custom fitted, leather covered liners and pads, and leather covered ear cups. These flight helmets not only looked great, they were also an improvement in comfort over the "government issue" flight helmets.

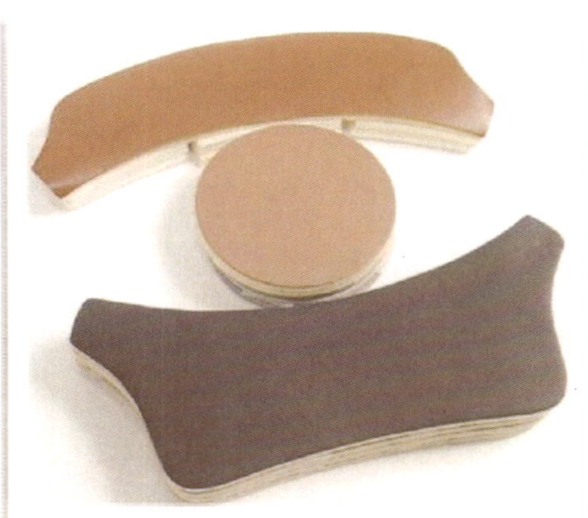

As with the TOPTEX helmets, we see the continued use of the Air Force Outstanding Unit Award, now with 3 oak leaf clusters. Another way of determining approximate era of undated Thunderbirds helmet photos is to examine the ribbons below the Thunderbird emblem.

The first Air Force Outstanding Unit Award was awarded in 1958, with several additional awards (oak leaf clusters) throughout the years. We first saw the inclusion of this ribbon on the TOPTEX helmets when the Thunderbirds were flying the F-100 "Super Saber." The first Air Force Organizational Excellence Award was not awarded until 1984, which provides another clue when attempting to determine the era of a specific helmet. Closer examination of oak leaf clusters (subsequent awards) makes it possible to further calculate the approximate date of a flight helmet.

Air Force Outstanding Unit Award

Air Force Organizational Excellence Award

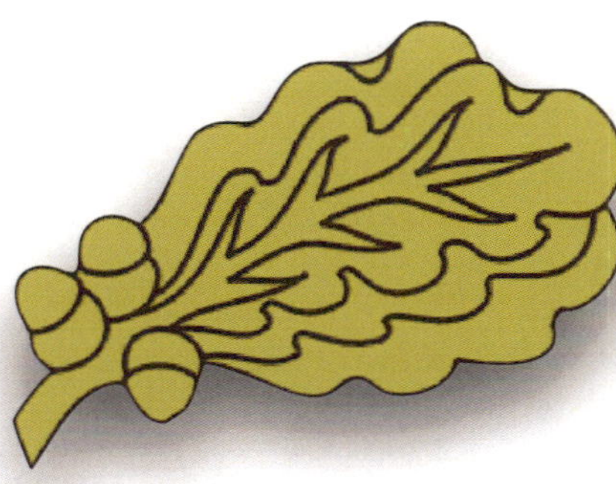

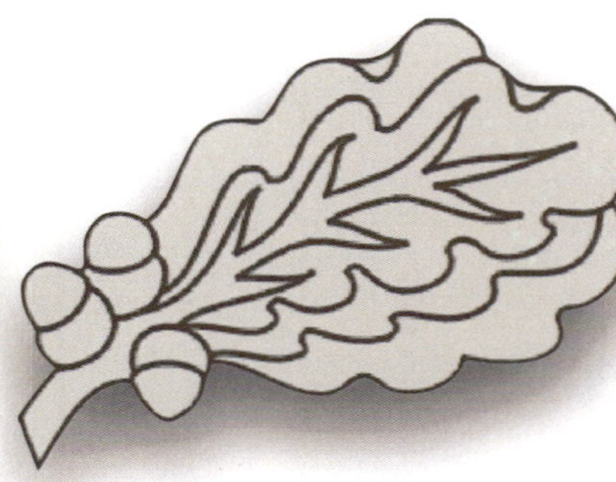

The bronze oak leaf clusters represent subsequent awards when attached to the ribbon, while the silver oak leaf clusters represent five (5) awards. The Thunderbirds' Air Force Outstanding Unit Awards and the Air Force Organizational Excellence Awards and award dates can be found in the Appendix section of this book.

By way of example, Gen. Lloyd "Fig" Newton's flight helmet bears an Air Force Outstanding Unit Award ribbon with three (3) oak leaf clusters:

The ribbon represents the first award, and the three (3) oak leaf clusters represent the second, third and fourth awards. (See image next page)

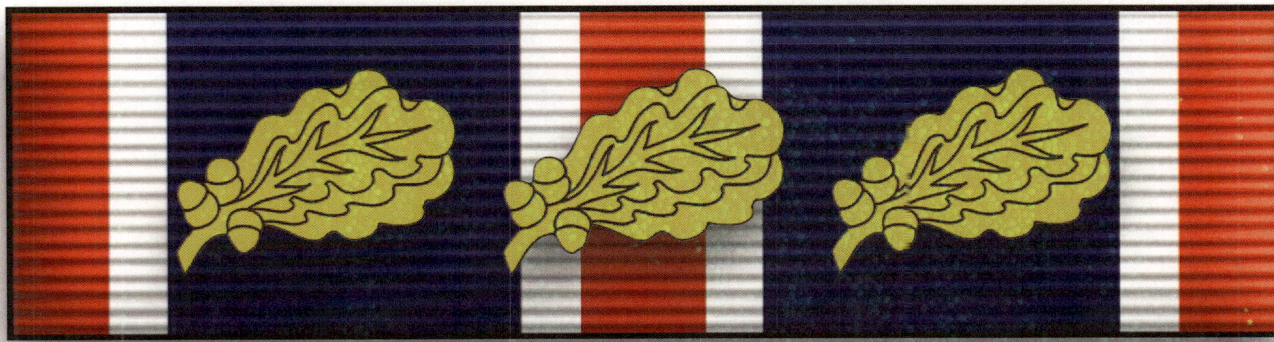

The inclusion of unit citation ribbons positioned below the Thunderbirds emblem would be discontinued around 2015[8].

In the early 1970s, the fuel crisis forced the Thunderbirds to select a more fuel efficient aircraft. They chose the USAF's supersonic trainer, the Northrop T-38A "Talon." It was the right choice since five (5) T-38s used the same amount of fuel as one F-4 Phantom, and required fewer people and equipment to maintain the aircraft. Although not a front line fighter, the sleek and graceful Talon served the team well from 1974 to 1982.

This was the era that I was first introduced to the Thunderbirds as an aerial demonstration team. It would be another 20 years before I would become obsessed with their flight helmets.

The aircraft had changed but, the helmets remained the same...with the exception of the 4-ship diamond formation silhouettes inside the Thunderbirds emblem.

This first of the T-38A "Talon" helmets we examine is a unique example in the private collection of John Slezak (next page, bottom photo). This shell is completely painted and graphics applied, but missing the edge roll and internal components. Of particular

[8] 2018 email from Gibson & Barnes citing Commander preference.

Photo Credit: John Slezak (Personal Collection)

interest and curiosity are the color of the stars, which are a lighter, brighter blue than the stars on similar helmets of the same era. It is unknown if this was an error, or, simply experimenting with something new. Whichever the case, this is the only example of this lighter blue star version found as of this writing.

It is also noteworthy that the Air Force Outstanding Unit Award ribbon has one (1) silver oak leaf cluster, representing six (6) awards.

The top photo compares and contrasts the F-4E "Phantom" alongside the T-38A "Talon" helmet with the lighter blue stars. In these photos, it is obvious that there are slight variations in the scallop design. The size of the helmet (Medium, Large, Extra Large) plays a part in this as well as the various artists who are laying out these designs by hand.

The photos below also illustrate some of the anomalies encountered when conducting research. The stock number printed on the white chin strap bears a manufacturer's date of 1980, placing this helmet in the late T-38A era; however, it is still equipped with white, cast aluminum oxygen mask receivers (for "T" bayonets) consistent with F-4E era aircraft or very early (1970s) T-38A aircraft. Also, the nape strap is black and the chin strap is white. One of many mysteries that make collecting and aviation history fascinating and challenging!

In an ongoing effort to document subtle changes, we see the first indications that the helmet chin straps

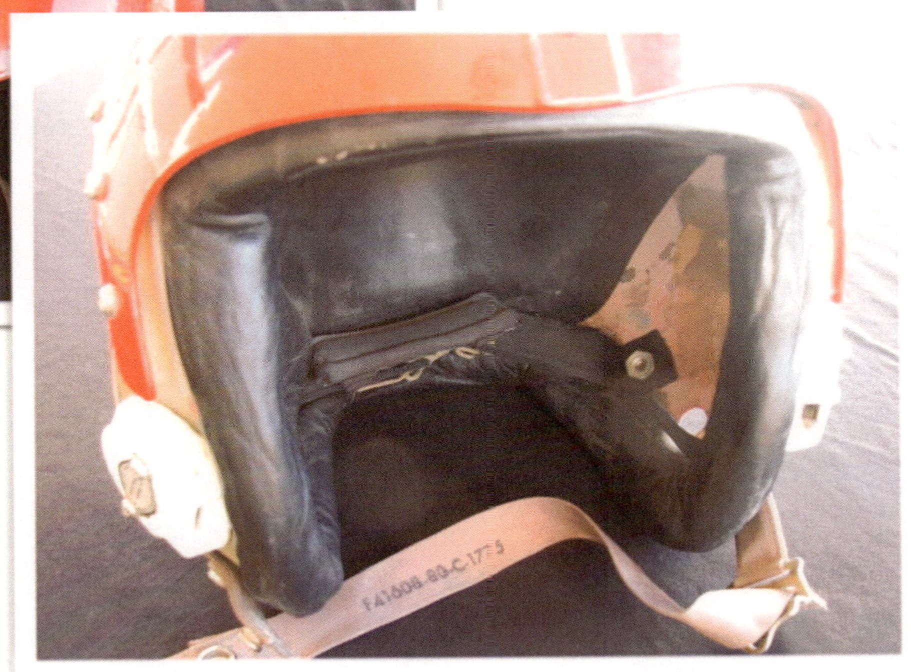

— once white — were now transitioning to black. The first photo on this page is of a T-38A era flight helmet from the 1979 - 1981 show seasons with a black chin strap. The other two photos are from Gen. Lloyd "Fig" Newton who served with the Thunderbirds from 1975 to 1978.

The photo of Gen. Newton on the ladder of his T-38A shows a helmet with a white chin strap, but the helmet he donated to the Smithsonian National Museum of African American History is fitted with a black chin strap. The difference may be due to two different helmets, or, the same helmet with an upgraded chin strap.

In just a few short years, the Thunderbirds would once

again change aircraft — and flight helmets, which would feature a permanent switch to all black chin and nape straps.

Pictured below is a comparison of the chin and nape straps pictured in white and black. Other than the color, and some leather upgrades, the basic design has remained unchanged for decades. Despite the longevity of this time-proven design, an innovative change, enhancing safety, would not take place until late in the second decade of the 21st century.

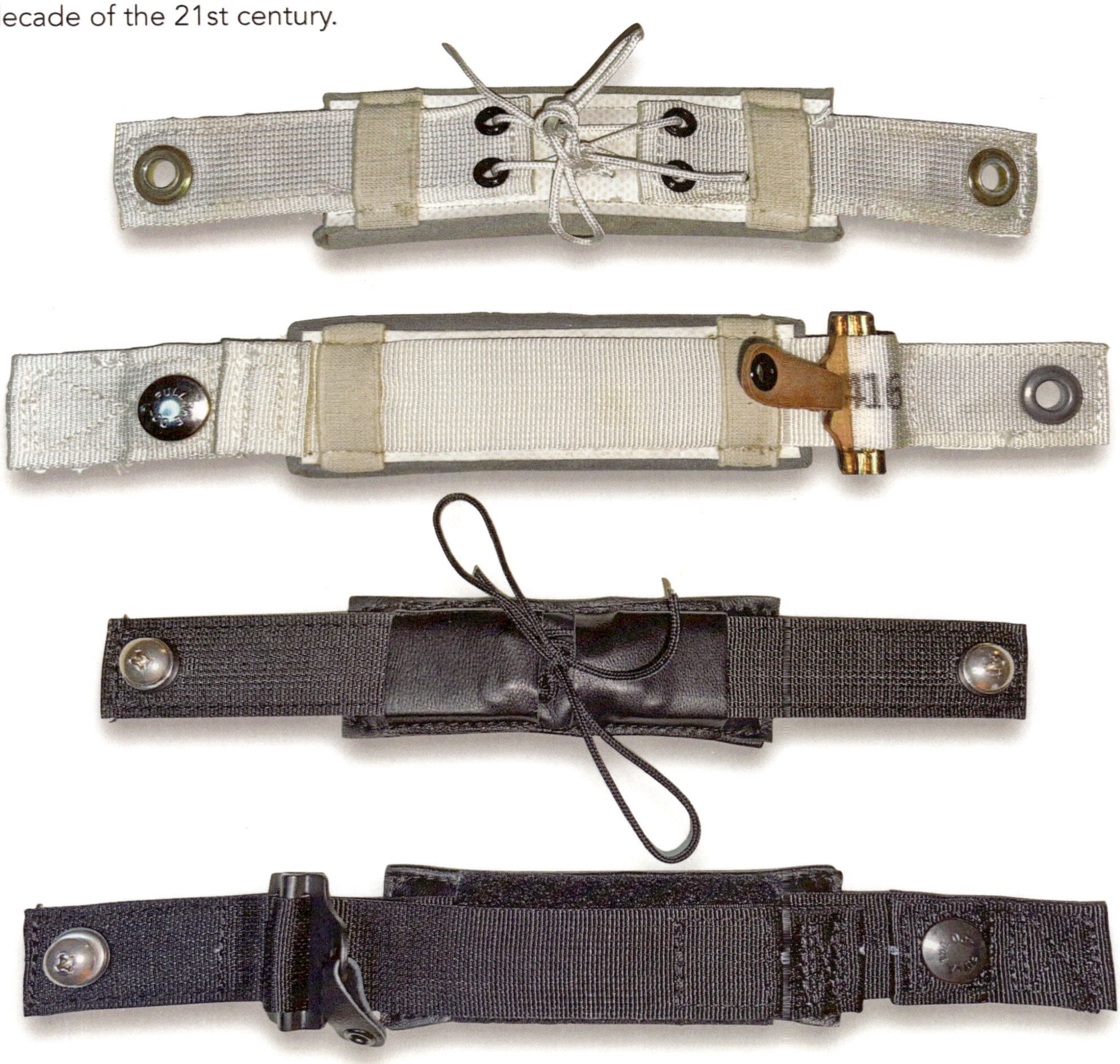

Brigadier General E. Dan Cherry[9]
USAF Thunderbirds Commander / Leader
1977 - 1978

USAF Photo

In the late summer of 2018, I wrote a letter to Brig. Gen. Dan Cherry, USAF (Ret'd) Thunderbirds Commander / Leader for the 1977 and 1978 show seasons requesting an interview for this book. I was honored and humbled that Gen. Cherry accepted. This is a rare opportunity to interview a man whose military biography reads like a "Who's Who" of the military fighter pilot community.

To say Gen. Cherry had a very distinguished career as a USAF Command Pilot would be a gross understatement. He logged 4,000+ flying hours, and served two tours in Vietnam, completing 110 combat missions in the F-105 "Thunderchief" (1967) and 185 combat missions in the F-4D "Phantom" (1971). He also was credited with the destruction of a North Vietnamese MIG-21.[10]

---

[9] Gen. Cherry was a Lt. Col. in 1977-1978.

[10] Gen. Cherry authored the book, "My Enemy, My Friend, a Story of Reconciliation From the Vietnam War"

Gen. Cherry's military decorations and awards include (but are not limited to) the Silver Star with oak leaf cluster, Legion of Merit with oak leaf cluster, Distinguished Flying Cross with nine (9) oak leaf clusters, Meritorious Service Medal with two (2) oak leaf clusters, Air Medal with thirty-four (34) oak leaf clusters, and Air Force Commendation Medal. [11]

Photo Credits: Gen. Dan Cherry

Gen. Cherry not only commanded the USAF Thunderbirds, but was commander when Gen. Lloyd "Fig" Newton, the first African American, also flew with the team.

My interview took place in mid September 2018 near the Western

[11] https://www.af.mil/About-Us/Biographies/Display/Article/105103/brigadier-general-e-daniel-cherry/

Kentucky University campus, Bowling Green, KY, a short 2 hour drive from my home. The date, time and place was set. I had carefully prepared my leading questions. My digital recorder was fully charged. This was truly going to be a once-in-a-lifetime, epic experience. It was time to meet the man behind the story.

It seemed appropriate, albeit coincidental, that this interview would take place on 18 September, the 71st birthday of the United States Air Force!

Gen. Cherry arrived promptly at the appointed time, sporting a white polo shirt with what appeared to be the Thunderbirds emblem embroidered on the left breast; however, upon closer examination, the script read, "Thundergeezers!" Gen. Cherry advised that his team mates from the 1977 and 1978 show seasons meet for a reunion every other year and have affectionately named themselves the "Thundergeezers" referencing their "senior" Thunderbird status! His personal vehicle was also adorned with a vanity license plate that read, "TBRD 1." Roger that, Sir!

Gen. Cherry shared that his passion for aviation began early in his childhood, influenced by his father, who was an aeronautical engineer in the 1940s and 1950s and, having always lived near an airport. His father worked on projects such as the B-49 "Flying Wing," the C-130 "Hercules," and later the C-5 "Starlifter." He also mentioned being inspired by an uncle, who was a WWII B-26 pilot. He was particularly enamored by the silver wings on his uncle's uniform, making a lasting impression on him during a time when military pilots were looked upon as heroes.

In 1959, Gen. Cherry enlisted in the United States Air Force Aviation Cadet Program[12] at James Connally Air Force Base.[13] This program was available to single (unmarried),

[12] Originally created in 1907 by the U.S. Army Signal Corps to train new pilots.

[13] James Connally AFB was closed in 1968 and located north of Waco, TX.

non-degreed applicants, who could successfully pass a barrage of physical and psychological testing. After completing an intensive one year program, Gen. Cherry earned his wings and a commission as a 2nd Lieutenant in the United States Air Force.

USAF Photo

Due to a shortage, Gen. Cherry initially trained as a navigator in the F-89 "Scorpion" and was assigned to an RC-121 squadron prior to acceptance in pilot training in 1964. Flight instruction was conducted in the T-37 "Tweet" and the T-33 "Shooting Star" and his first assignment as a new pilot was in the F-105 "Thunderchief' at Nellis AFB, NV.

With the war in Vietnam ramping up, Gen. Cherry volunteered for duty, flying 110 combat sorties over North Vietnam in the F-105 in 7 months! Upon rotation back to the United States, he became an instructor in the F-105 at McConnell AFB from 1968 to 1971 before transitioning to the F-4D "Phantom." With tensions continuing to escalate in Vietnam, Gen. Cherry volunteered for a second tour where he flew an additional 185 combat sorties over North Vietnam, scoring a MiG-21 shoot down[14].

"I would have never been selected as a Thunderbird had I not served a second tour in Vietnam."

Brig. General Dan Cherry

Having served two tours tours in Vietnam, combined with his numerous awards and decorations, Gen. Cherry said he had credentials that very few people had at the time

[14] Gen. Cherry's Weapons Systems Officer (WSO) was Jeffrey Feinstein, 1 of only 5 Vietnam ACES

and this set him apart from his peers. He said that Thunderbird assignments are very competitive, adding that without the second tour in Vietnam, he would have never been selected as a Thunderbird commander.

I was in awe during this interview and felt I was in the presence of a fighter pilot's fighter pilot — an aviation rock star, yet he spoke softly and very nonchalantly. It is extremely rare to meet a man of this caliber and with these accomplishments who comes across this humble and reserved — a true gentleman in every sense of the word. They don't make them like this anymore.

When we got into the details of his service with the Thunderbirds, Gen. Cherry said he always aspired to be a Thunderbird, and as a Captain, applied for a team position, but was not selected. Following his Vietnam service tours, he was considered "too old" or "too senior" for a wing or solo position, but had the qualifications for Commander and Leader, so sent in his application package. He was excited when he was notified that he was a "semi-finalist" along with a handful of other guys all vying for this one cherished position. There had only been 22 Thunderbirds Commanders / Leaders prior to this selection process. Indeed a rare breed.

These "semi-finalists" were invited to join the Thunderbirds team on the road during a scheduled air show weekend to get a "dose" of what this fast-paced life was like with a world class flight demonstration squadron. This group was further whittled down to a "finalist" list of 4 or 5 candidates who were invited, along with their spouses, to Nellis AFB, home of the USAF Thunderbirds. While there, the candidates would be subjected to interviews, flying evaluations in the T-38 "Talon" and social gatherings with the current team, candidates and spouses. Gen. Cherry pointed out that these social gatherings were very much a part of the interview process to allow team members an opportunity to witness the candidates in a social setting, which is critical to the Thunderbirds' Public Affairs mission. In addition, the "chemistry" had to be right with the new commander. He had to fit in as well as be respected by the team.

Despite never having flown a T-38, Gen. Cherry flew with a team evaluator not only to review his "stick" skills, but observe his judgement and decision making process while flying in formation with other T-38 aircraft. He said he liked the T-38 and it handled similarly to the F-105, his all-time favorite aircraft.

The candidates then returned to their duty assignments and waited for a "the call." The team gets together and makes recommendations on their choice for the new commander, but the final decision is made by the Tactical Air Command (TAC)[15] Commander, a 4-star general. Historically, the TAC Commander would typically "rubber stamp" the team's recommendation, respecting their judgement.

> **"Dan, the decision has been made on our new Commander / Leader, and we want you to join the Team."**
>
> Major Chris Patterakis
> USAF Thunderbirds Commander / Leader
> 1975 - 1976

Gen. Cherry remembers he was stationed at the Pentagon during the selection process in the Fall of 1976 and living in Northern Virginia. He was home over the weekend and his parents happened to be visiting. The family was gathered in the living room enjoying a cocktail when the phone rang. It was Major Chris Patterakis — the current Thunderbird Commander, calling. Gen. Cherry said, "My heart almost stopped!"

Gen. Cherry knew this call was news about the new Commander / Leader, but he feared the worst. Maj. Patterakis reportedly said, "Dan, the decision has been made on our new Commander / Leader, and we want you to join the team."

Gen. Cherry said of being selected, "It was an awesome experience…you've got a tiger by the tail!" Within a few weeks, he had orders to Holloman AFB, where they trained aggressor squadrons. This is where Gen. Cherry received his checkout in the T-38 for 6 weeks, before reporting to Nellis AFB.

---

15 Tactical Air Command is now Air Combat Command. The current selection process has changed.

During this time, Thunderbird pilots went to Wright-Patterson AFB in Dayton, OH to be fitted for their flight suits.[1] Gen. Cherry stated that they were issued several flight suits in red, white, blue, grey and black.

When asked about anti-g suits, he replied, "In my day, we didn't wear g-suits...just didn't need them." He added that in the T-38 they experienced limited G's and were in good physical condition. Besides, "It screwed up our show suit — it made it look bad, and we didn't need them."

Photo credit: USAF Thunderbirds

Upon reporting to Nellis AFB, he received instruction from Maj. Chris Patterakis on how to be a Commander / Leader of the USAF Thunderbirds. Gen. Cherry mentioned a very specific syllabus of training, spending time with Thunderbird #8 "shadowing" the Leader, in order to learn the routine. On occasion, he would fly in the back seat of #1 where he could experience the position first hand.

Regarding receiving his Thunderbird flight helmet, Gen. Cherry said that this was very much a "treasure" and a "badge of honor" — a memory he cherished. The only personalization was the small "Dan" hand-painted at the base of the visor housing track. This brought up the question of call signs, to which Gen. Cherry advised he never had a call sign except with the Thunderbirds — "Boss" — which is what they call him today.
We briefly discussed low points as a Thunderbird, and as low points go, it is a commander's worst nightmare: the loss of a team member.

[1] The flight suits are now tailored by Flight Suits Ltd., dba Gibson & Barnes of El Cajon, CA, a third-party supplier.

On July 25, 1977, Captain Charlie Carter, Thunderbirds Narrator from San Antonio, TX aborted a landing at Cheyenne Regional Airport due to wet and slippery conditions. Gen. Cherry said the tires are small and the T-38 is not equipped with a drag chute, and approach speeds are high, making landing on a wet runway very challenging.

As Capt. Carter was attempting to gain altitude, the aircraft crashed through a campground at the departure end of the runway. Capt. Carter ejected from the aircraft, but did not survive. Staff Sargent Ted Foster (Crew Chief) of Charlotte, North Carolina, was in the rear seat and ejected successfully, sustaining only minor injuries. The danger is ever present. The pain is real.

"The danger is ever present.
The pain is real."

At the conclusion of my interview, Gen. Cherry invited me to visit the Western Kentucky University Museum in Bowling Green, Kentucky, where his Thunderbirds flight helmet from the 1977 and 1978 show seasons is on loan. We were graciously allowed to photograph this beautiful work of art customized by Protection, Inc.

Photos's of Brig. Gen. Dan Cherry's HGU-26/P Flight Helmet custom painted by Protection, Incorporated and MBU-5/P Oxygen Mask used in the 1977 and 1978 show seasons. Photos by author at the Western Kentucky University Museum, Bowling Green, KY.

Comparing and contrasting three T-38-era Thunderbird flight helmets. TOP: Daniel G. Wilbur, Photographer, 1979 - 1981; CENTER: Boss Cherry, Commander / Leader, 1977 -1978; BOTTOM: SSgt. Mike Turkington, Life Support, 1979 - 1981.

The Red Helmet

This flight helmet belongs to Brigadier General Jim Latham, USAF (Retired) who flew Right Wing from November 1978 to January 1981. (Photo credit: Brig. Gen. Latham)

6

HGU-55/P & HGU-55/E Series Flight Helmets

According to GENTEX, the HGU-55/P helmet assembly was provided to USAF test & evaluation units in 1981, as a replacement for the HGU-26/P-series helmet. First major deliveries to squadrons began showing up in 1983.

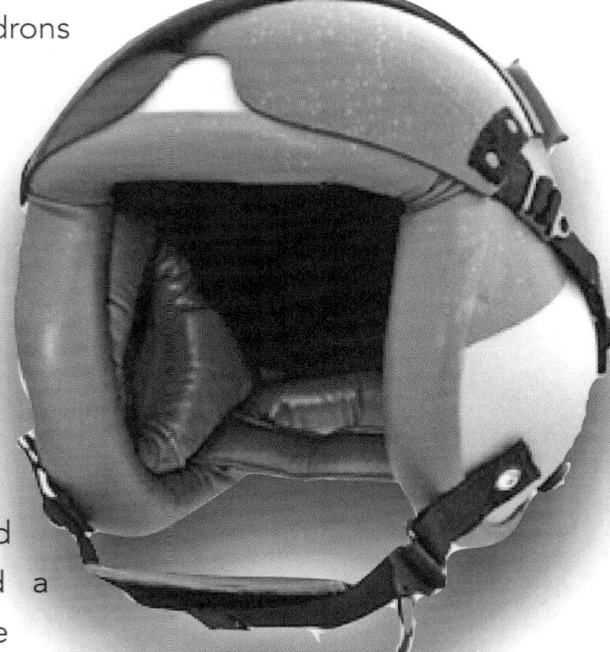

HGU-55/P Flight Helmet with Lightweight External Visor. (GENTEX Photo)

The shell is typically made of fiberglass fabric[1] molded with epoxy resin. It is equipped with an energy-absorbing styrofoam liner and a custom fitted Thermoplastic liner (TPL). The helmet is equipped with a single snap-on visor that eliminates the visor housing for reduced weight and a lower profile but it is capable of accepting the PRU-36/P dual visor assembly or an EEK-4A/P single visor housing assembly. The helmet shell is cut back at the temples to afford the wearer with increased peripheral vision capability. The HGU-55/P integrates the H-154A/AIC Headset and lightweight oxygen mask receiver kit. This helmet was normally utilized with USAF standard MBU-5/P, MBU-12/P oxygen masks and later, MBU-20/P.

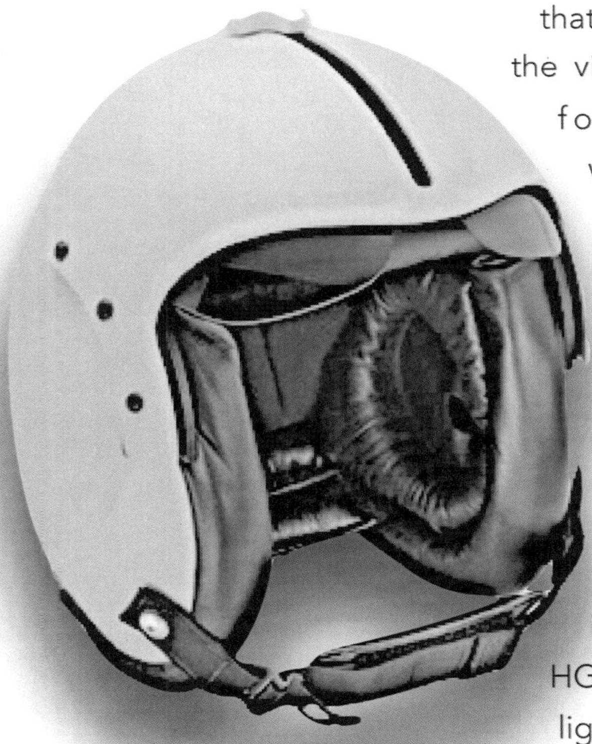

HGU-55/E Flight Helmet With EEK-4A/P Single Visor Housing (GENTEX Photo).

[1] Also, Kevlar.

The HGU-55/E[2] is identical to the standard HGU-55/P except that a protective visor assembly — designated EEK-4A/P — replaces the bungee visor. Its protective visor housing allows easy movement of the impact-resistant polycarbonate lens and locks the lens securely up or down.

Despite the increased weight, the Thunderbirds chose the "E" configuration over the "P" version to allow them to continue their long standing tradition of adorning their helmets and visor housings with visually stimulating graphics, aircraft numbers, and personalized markings such as names and call signs.

The HGU-55/P came standard with a flat grey painted shell, a grey leather edge roll and two "elephant ear" leather pads attached at each temple to provide scratch protection from the bungee visor straps. These "elephant ears" as they were referred to, were not used on the Thunderbirds version, and the leather edge roll is black instead of grey.

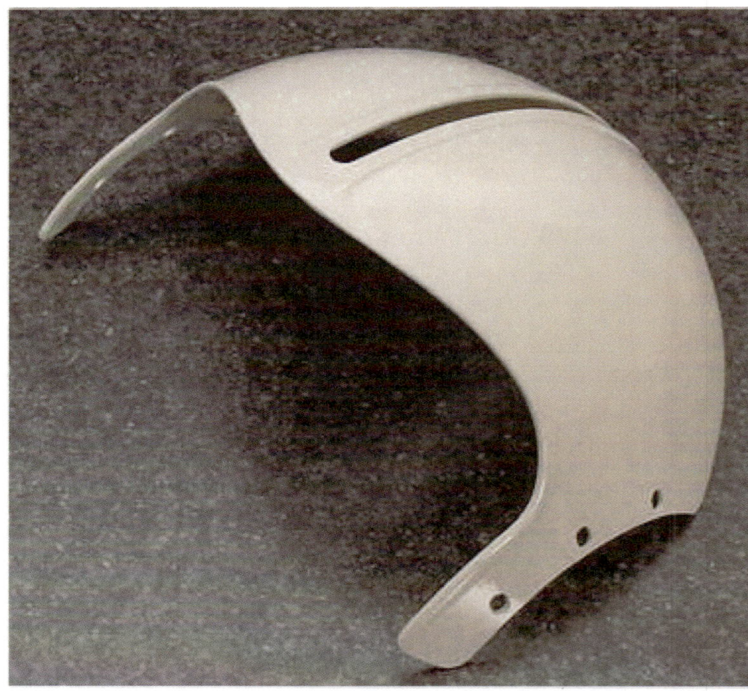

EEK-4A/P Single Visor Housing to convert a flight helmet from an HGU-55/P to an HGU-55/E

The lightweight bayonet receivers were also exchanged for chrome plated, cast aluminum receivers, as was the choice in previous helmet versions, which increased the weight, but are more aesthetically pleasing.

2 The "E" designation is Gibson & Barnes nomenclature, not GENTEX

Lightweight oxygen mask bayonet receivers as used on the HGU-55/P flight helmet. *(Author's photo)*

Chrome plated oxygen mask bayonet receivers as used on the Thunderbirds HGU-55/P flight helmet. *(Author's photo)*

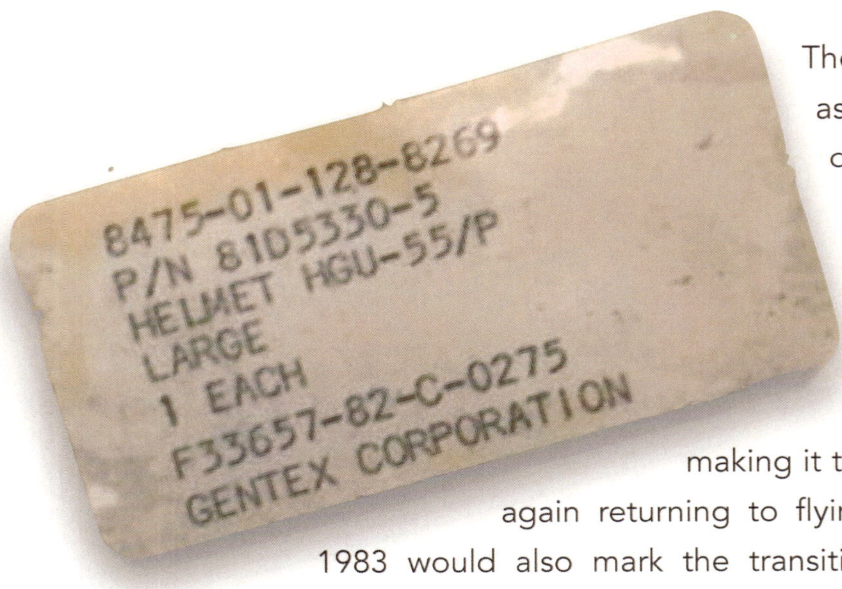

The first red, white and blue F-16A assigned to the Thunderbirds was delivered to Nellis AFB on Jun. 22, 1982. Due to the conversion to the new aircraft, there were no official shows flown in 1982. The team first flew the F-16 during the 1983 show season, making it the team's eighth aircraft and once again returning to flying a front-line fighter.[3] The year 1983 would also mark the transition to the new HGU-55/E flight helmets.

When the Team upgraded to the new HGU-55/E flight helmet, it would be difficult to the untrained eye that a change had been made. The most notable difference is the new profile at the temples, which are "cut back" as compared to the HGU-2A/P and HGU-26/P series helmets. The addition of the EEK-4A/P single visor housing tends to further camouflage this change, and can only be verified upon close examination.

[3] http://afthunderbirds.com/site/history/

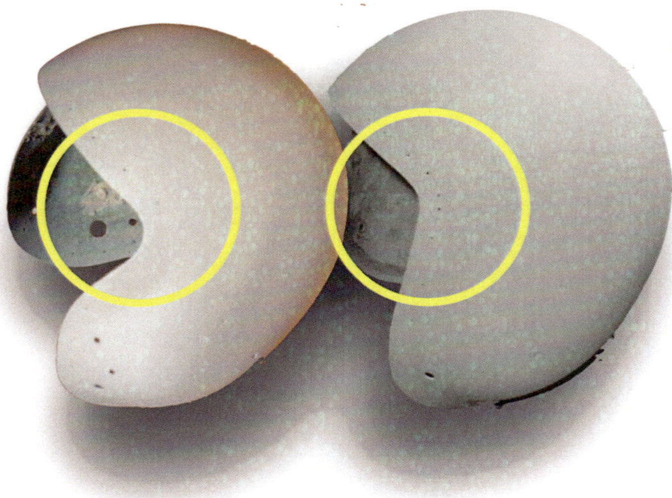

The two photos on the left illustrate the differences in the trimming of the temples. The photo on the far left is an HGU-55/P shell with a more pronounced "cutback" than the HGU-26/P version on the right.

The photo below is Brig. Gen. Jim Latham,[4] Commander / Leader of the Thunderbirds for the 1982 to 1983 show seasons. Proudly standing on the ladder of his F-16A "Fighting Falcon," Gen. Latham sports a rare white flight suit and his new HGU-55/E flight helmet. Gen. Latham was the first Commander / Leader in the F-16A "Fighting Falcon."

USAF Photo

Gen. Latham also served as Right Wing in the T-38 from November 1978 to November 1979, as well as Operations Officer and Right Wing from November 1979 to January 1981.

Gen. Latham returned to the Thunderbirds as Commander / Leader following the tragic Indian Springs crash in January of 1982 that took the lives of four Thunderbird pilots.

[4] Major at the time of service

I was fortunate to locate Gen. Latham and began an exchange of email messages that I parlayed into a telephone interview in February 2019.

Gen. Latham was selected, and accepted the challenge to accomplish two incredibly daunting missions: 1.) Rebuild the Thunderbirds, and 2.) Transition the Team from T-38s to F-16s. It's difficult enough to imagine that, individually, the jobs would require his full, undivided attention; together, the devotion necessary to succeed would seem overwhelmingly impossible by most.

According to Gen. Latham, 3 to 4 months following the Indian Springs crash, the USAF Thunderbirds, the U. S. Navy Blue Angels and the U. S. Army Golden Knights implemented a safety "stand down," as their collective futures underwent intense scrutiny through a risk vs. reward analysis. General Wilbur L. "Bill" Creech, then Commander of Tactical Air Command[5] and former USAF Thunderbird pilot[6], made a compelling argument to the Secretary of the Air Force[7] who in turn convinced President Ronald Reagan of the importance, commitment and value the Thunderbirds add to the Air Force's recruiting and retention efforts. In April of 1982, the decision was made to rebuild the Team, with Secretary Verne Orr announcing, "The Thunderbirds will fly again."

> **"The Thunderbirds will fly again."**
>
> Verne Orr
> Secretary of the Air Force
> 1981 - 1985

Gen. Latham joined the Team as Commander / Leader in June 1982, and spent the remainder of the calendar year building the new team with some members who

---

[5] Tactical Air Command now incorporated into Air Combat Command.

[6] 1954 and 1955 show seasons.

[7] Verne Orr, 1981-1985

previously selected for T-38s, and new members, who were from the F-16 and F-15 communities. As you might have expected, no air shows were performed in 1982.

Gen. Latham didn't recall a great deal of detail about his Thunderbird flight helmets, as he had more pressing issues on his mind at this time; however, he did remember there was a slight difference between the helmet he wore in the T-38 (HGU-26/P) and the one he wore in the F-16 (HGU-55/P), the most notable difference was lighter in weight. He graciously agreed to provide photographs of his helmet from the F-16 years, shown below.

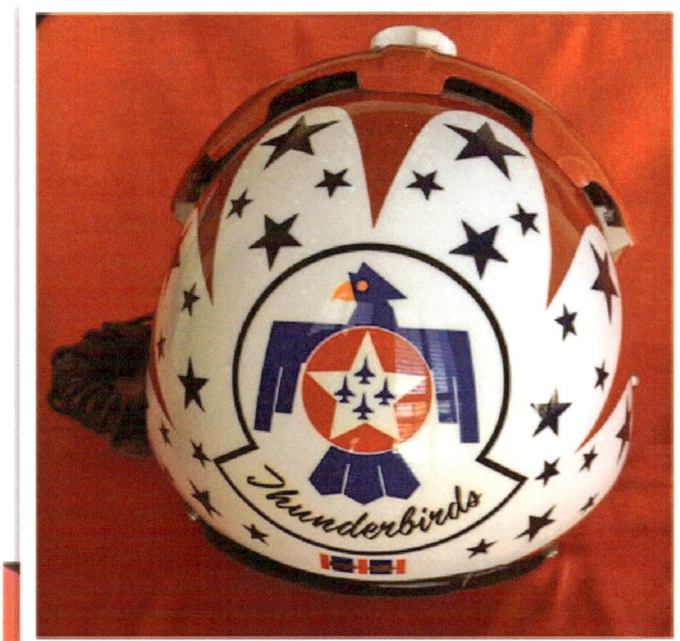

We have further proof that the era is accurate based on the fact that this helmet does not have an Air Force Organizational Excellence Award ribbon, which was not awarded to the Thunderbirds until 1984.

As of this writing, the HGU-55/P/E flight helmet has been the longest in service with the Thunderbirds — 36+ years — with minimal upgrades and modifications along the way. This helmet also saw the addition of USAF logos, aircraft numbers, pilot names and call signs added to the visor housings, making these the most personalized in Thunderbirds history. These graphics, fonts and call signs provide us with clues to further determine the changes that occurred through the evolution of this magnificent flight helmet.

In the three examples below, it is difficult to tell the difference between these HGU-55/E helmets and their HGU-26/P predecessors; however, we do see the addition of the Air Force Organizational Excellence Award, first issued in 1984.

Another noticeable difference is the layout of the graphics. I suspect the helmet on the left and center (below) are size LARGE, and the helmet on the right is an extra large. The Thunderbird emblem remains a constant size, so it appears smaller on a larger "canvas." The two small stars on either side of the ribbons have also been eliminated, only to reappear — and disappear again —- in subsequent examples.

In my many conversations with the Thunderbirds Account Manager at Flight Suits, Ltd. in the mid-1990s, and more recent conversations with the Thunderbirds Account

Manager at Gibson & Barnes in 2018, I was told these minor changes to include — or exclude — two small stars flanking the ribbons are directed by the Thunderbirds, and at the discretion of the commander. Accurately tracking and dating these changes is difficult, if not impossible!

Research for this project began in the mid-1990s and involved an untold number of phone calls to the Thunderbirds Life Support Shop[8] at Nellis AFB, NV. The shop is staffed by some very knowledgable, professional and helpful airmen who were always eager to take my calls and answer my seemingly never ending questions. In support of my research, a couple of the Life Support technicians agreed to meet me at an air show — The Sertoma Cajun Air Festival — in Lafayette, LA in May of 1996 and allow me to take photographs of a spare Thunderbirds helmet.[9] This was the first "authentic" Thunderbirds helmet I had the opportunity to examine up close and personal - I couldn't contain my excitement!

These photos were very valuable over the years as I continued my pursuit of comparing and contrasting the evolution and many variations of this elegant flight helmet.

[8] Now called Aircrew Flight Equipment (AFE)

[9] The author's Thunderbird Flight Helmet Project is detailed in a subsequent chapter.

Throughout the years, the black border that frames the Thunderbirds emblem has also changed periodically from thin to thick to thin, with no apparent rhyme or reason! The examples on this page clearly show the variations in border thickness.

Sometime in the 1990s, Thunderbirds Life Support began to add aircraft numbers and pilot names or "call signs" on the visor housings. The earliest examples were simple: Name or call sign on the wearer's right, and the aircraft number on the wearer's left (Top left and center photos).

By the early 2000s, the name and the aircraft number were both located on the wearer's left, and the new USAF emblem was added on the right side. This visor housing (left) is from Major Todd Canterbury who flew with the Thunderbirds between 2002 - 2004 (Bottom left).

In the 2006 - 2007 show season, the font and style appear to be unchanged; however, as the two examples on the next page illustrate, pilots could - and would - periodically change things up. Major Nicole Malachowski was the first female

USAF Thunderbirds pilot, flying the #3 (Right Wing) jet. These examples illustrate that pilots could — and would — periodically switch between their name and call sign on their visor housing.

The 2007 - 2008 show season witnessed the second female pilot is as many years to join the prestigious ranks of the Thunderbirds. Major Samantha "Combo" Weeks flew the #6 Opposing Solo position in 2007 followed by the #5 Lead Solo position in 2008. The photo on the next page clearly shows a black and white "shadow" added to the USAF emblem as well as to her call sign, "Combo."

The font has also been changed. The inset photo is a helmet belonging to Captain Kristin "Mother" Hubbard who filled the Advance Pilot and Narrator positions in the 2010 - 2011 seasons.

Another change that's difficult to pinpoint, but is reflected in the photos above and on this and the previous page are the discontinued use of custom logos on the oxygen mask receiver "diamonds."

The first photo below is from Flight Suits, Ltd. (FSL) and can be dated between 1983 and the 1990s. The second one, also from FSL, dates between the 1990s and 2000s. The third example is from Gibson & Barnes and was introduced in the 2000s. The fourth example is blank - the diamond is turned inside-out. We see this change begin to occur around the early to mid 2000s. When I spoke with Gibson & Barnes, they advised this was done by the Thunderbirds and they were not aware of this change. I later learned from the Thunderbirds Public Affairs that this decision was based on avoiding any optics of sponsorship or endorsement.

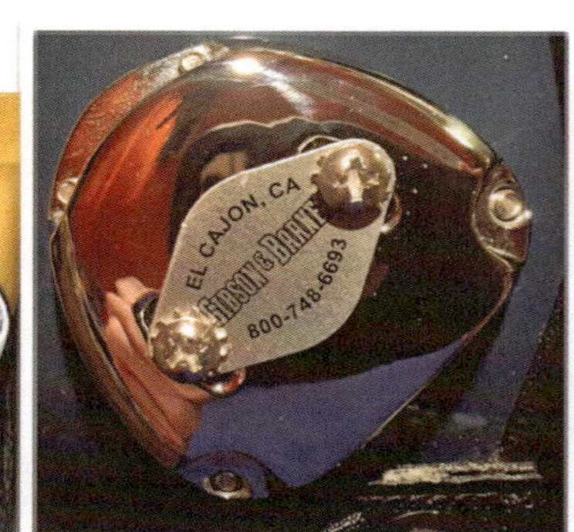

The next font change appears to have entered somewhere around the 2013 - 2015 show seasons, with what appears to be a "deeper" shadow. The helmet on the top left belonged to Major Joshua "Tater" Boudreaux who flew the #2 Left Wing position. The inset photo is Major Jason "Jaws" Curtis' helmet from the same 2013 - 2015 show seasons.

The next font changes came along around the 2015 show season with a bolder, deeper shadow with distinctive white highlights. The photos on the next page reflect this unusual font.

The first significant change to the HGU-55/P came in the early 2000-teens with the introduction of the integrated chin and nape strap. Prior to this, the chin and nape straps were separate, individual straps requiring adjustments to both to properly fit the wearer.

The top photo (next page) is of Major Scott "Cheetah" Petz, who served as the Advance Pilot / Narrator for the 2015 - 2016 show seasons. The center photo is Major Nate "Hassel" Hofmann, Right Wing, 2017 - 2018 show seasons, and the bottom photo is Captain William "Boar" Graeff, Left Wing, 2018 - 2019 show seasons.

The Integrated Chin and Nape is a unique, one piece assembly threaded through the helmet and adjusts at the chinstrap. Cross straps threaded through the nape strap pad adjust simultaneously with the chinstrap to provide a snug yet comfortable fit ensuring helmet stability and retention

The chin / nape assembly is made of nylon webbing and features chin and nape strap pads covered with leather. Sliding clamps are installed on the left and right chin straps at the shell slots to allow individual adjustment of the straps if desired.[10]

(USAF Photos)

10 Gibson-Barnes.com

Individual Chin & Nape Straps
Integrated Chin & Nape Strap

HGU-55/P Individual Chin and Nape Strap Assembly (Top and Bottom)

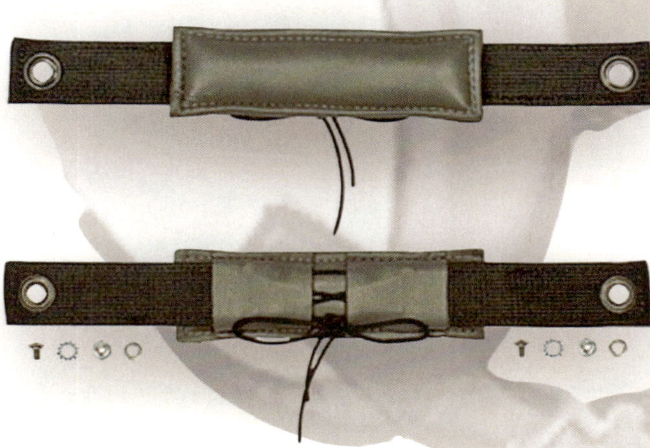

HGU-55/P Integrated Chin and Nape Strap Assembly (Top and Bottom)

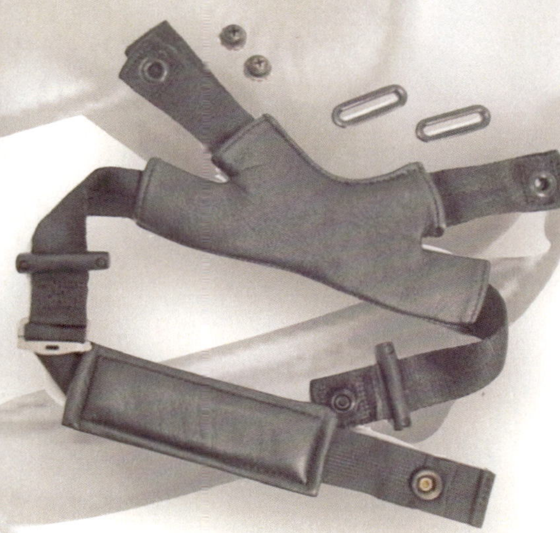

The photos on this page all have unusual or non-standard graphics on the visor housings.

I reached out to TSgt. Jasper Roberts, USAF Thunderbirds Aircrew Flight Equipment (AFE) Shop at Nellis AFB for answers.

According to TSgt. Roberts, the visor housings with the dual numbers are seen during winter training when new pilots are assimilating into the team. New pilots trained for their position will have a #2, which identifies the pilot as the incoming pilot, while the instructor pilot will have a #3, identifying him or her as the pilot instructor. The second number is the position being trained for by the student, and the number being vacated by the instructor. In the example above, "Jaws" was the instructor pilot for the #5 position. His student (incoming pilot) would wear the numbers 2 and 5. This is done as long as two pilots (student and instructor) are occupying the same aircraft. The practice of single numbers returns when the student graduates to solo status with the Team, and the instructor vacates the Team.

The photo on the right indicates that "Khan" is an instructor pilot for the #4 slot position.

In the 1990s and into the 2000s, you didn't have to be a Thunderbird or the 57th Wing Commander to get your own Thunderbird flight helmet.

During that time, Flight Suits, Ltd., now doing business as Gibson & Barnes offered them for sale to the public; however, they commanded a price tag of about $1,000…about twice the price today.

The photos on this page are courtesy of David Mattsson, a friend of mine and fellow enthusiast who ordered this stunning beauty in 2000, for his personal flight gear collection. By the way, the MBU-12/P oxygen mask is sold separately.

Today, that practice has changed, and "civilians" or "non-Thunderbirds" are required to request and obtain permission from the current Thunderbird Commander / Leader to get one of these cherished flight helmets.

The photo on the left (top) is another beautiful example of a customer made Thunderbirds flight helmet, courtesy of Alan Crane, a collector colleague in Australia.

Alan purchased this helmet from Flight Suits, Ltd. in 1993, when the black border on the Thunderbird emblem was bolder and more pronounced.

The example on the right comes from a 2019 eBay listing. The description indicated it was purchased from Flight Suits, Ltd. in 1995. This example, although close in age to the two helmets on the previous page, used the thinner black border. Although the red paint appears noticeably different on this, and previous examples, I am certain it is the result of different lighting and cameras, not due to paint color changes.

For comparison purposes, the photo on the right is an authentic Thunderbirds flight helmet from the same era. This helmet belonged to Capt. Mike "Mo" Beale[11], Solo pilot, 1990 and 1991 show seasons.

The photo on the left — Capt. "Mo" Beale — comes from the 1990 USAF Thunderbirds annual yearbook. All officers were photographed using the same flight helmet, which is slightly different (thinner Thunderbird emblem outline) from the one he was issued and flew with while on the team. The helmet used for the yearbook photos was likely a spare, or one used for VIP flights, as it was easier than each officer having to lug their flight helmet to the photo shoot.

[11] Colonel Michael O. Beale, USAF (Retired)

The photo below is of Brig. Gen. Robert G. "Blend'r" Novotny, the current 57th Wing Commander, Nellis Air Force Base, which is the parent organization of the USAF Thunderbirds. Instead of an aircraft number above his call sign, he sports a single 5-point star, representing his military grade.

The bottom photo is of the previous 57th Wing Commander, Brig. Gen. Jeannie "Tally" Leavitt, who is getting "chauffeured" by Maj. Nate "Hassel" Hoffman. Gen. Leavitt is the first female U.S. Air Force fighter pilot, and a combat veteran in the F-15 "Eagle."

7

Combat Edge Flight Helmet

The end of the 2017 show season and the beginning of the 2018 show season were marked by unprecedented change and devastating tragedy. In November of 2017, The Air Force removed the commanding officer of the Thunderbirds demonstration team due to a loss of confidence in his abilities to lead the squadron.

Lt. Col. Jason "Shifty" Heard was dismissed as Commander / Leader of the Thunderbirds and pilot of the team's No. 1 jet. Brig. Gen. Jeannie "Tally" Leavitt, commander of the service's 57th Wing[1], dismissed Heard, citing a loss of confidence in his "leadership and risk management style."

"This was an incredibly difficult decision to make but one that is ultimately in the best interests of the Thunderbird team," she said in an official statement. "I am personally grateful for Jason's dedication to the 2017 season."

On Nov. 4, the Thunderbirds completed their 2017 season which included more than 70 air shows. The team previously cancelled a scheduled air show in June following the crash of one of the unit's jets during training, prior to a two-day performance at the Dayton Air Show in Ohio.

An Air Force investigation determined the crash was avoidable citing the F-16 was traveling too fast and had insufficient stopping distance as it landed on a rain-soaked runway before it overturned, destroying the $29 million aircraft. The board found the pilot failed to follow proper braking procedures and his vision was impaired by rainwater on the cockpit's canopy. The pilot and his passenger recovered.

A spokesman for the Thunderbirds, said the crash was not the reason Leavitt dismissed Heard, stating Leavitt had concerns that Heard's leadership style was resulting in

[1] The Thunderbirds are assigned to the USAF 57th Wing, Nellis AFB, NV

unnecessary risk within the Thunderbirds' demonstrations, which eroded the team dynamic.

"We are on the road together more than 200 days per year, executing flying operations with absolutely no margin for error. As a result, absolute trust and teamwork in both our professional and personal dynamics are foundational to our mission."

USAF Photo

Leavitt determined the team was better off with new leadership as it begins preparing for its 2018 schedule. As a result, Lt. Col. Kevin "Löwen" Walsh, who had served as the Thunderbird's operations officer since 2016, was named commander of the unit.

Unfortunately, change and tragedy continued early into the 2018 show season. On 04 April 2018, Major Stephen "Cajun" Del Bagno, #4 Slot Pilot tragically lost his life when his plane went down on the Nevada Test and Training Range during what the Air Force called "a routine aerial demonstration training flight."[2] Maj. Del Bagno is one of 20 Thunderbirds who have paid the ultimate sacrifice while serving as an "Ambassador in Blue."

[2] This book is dedicated to Major Stephen "Cajun" Del Bagno, who was the first F-35A "Lightning II" pilot selected to the Thunderbirds.

"We are mourning the loss of Major Del Bagno...He was an integral part of our team and our hearts are heavy with his loss."

Brig. Gen. Jeannie "Tally" Leavitt
57th Wing Commander

"We remember Cajun as an airman, a warrior, a talented fighter pilot, and a great friend ... He lived to fly and inspire the next generation."

Lt. Col. Kevin "Löwen" Walsh
Commander / Leader

Following a three-week safety stand down, and the cancellation of several scheduled air shows, the decision was made to push forward with the 2018 show season. Major Nick "Khan" Krajicek, 2016 - 2017 Slot Pilot was selected to return to the team due to his most recent experience in this position and availability from his current assignment.

Flight helmet graphic:
Major Stephen "Cajun" Del Bagno, #4 Slot

As a pilot, and a Cajun[3] myself (by heritage), I felt a personal connection and sense of loss as a result of this tragic accident, despite never having the privilege of meeting Maj. Del Bagno. I am grateful, and consider myself fortunate that I never experienced a serious inflight emergency, but have had

---

[3] Cajuns are an ethnic group mainly living in Louisiana consisting in part of the descendants of the original Acadian exiles—French-speakers from Acadia (L'Acadie) in what are now the Maritime Provinces of Eastern Canada.

some "close calls" that became valuable life lessons. Ever since my first flight in 1974, I was introduced to, and have never forgotten a famous anonymous quote:

"Aviation in itself is not inherently dangerous, but to an even greater degree than the sea, it is terribly unforgiving of any carelessness, incapacity or neglect."

Shortly after the accident, photos began to emerge on the Internet of the 2018 team members wearing a flight helmet with a modification referred to as "Combat Edge." The exterior changes are very subtle, but savvy collectors immediately noticed the difference, and questions were being raised about this change

The Combined Advanced Technology Enhanced Design G-Ensemble (COMBAT EDGE) is a pressure breathing for "G" (PBG) system that improves a high-performance fighter pilot's tolerance to high-G maneuvers and is

HGU-55/P Combat Edge Helmet Modification Kit KMU-511/P

effective in preventing G-induced loss of consciousness (GLOC). Items of the system include a G-suit and Counter-pressure vest, Integrated Terminal Block pressure regulator, modified HGU-55/P with KMU-511/P kit and MBU-20/P low profile oxygen mask.

The complete Helmet Modification Kit KMU-511/P, designed to be easily retrofitted into the standard HGU-55/P, consist of a urethane-coated nylon bladder, PBG feed tube and quick-disconnect connector, specially contoured polycarbonate visor, and hook and pile fasteners. This system provides automatic mask tensioning during PBG to keep the mask in position under pressure, but only at high G so the pilot does not have to wear the mask uncomfortably firm at all times.[4] (USAF Photos)

Most of this modification kit (yellow bladder) is installed on the inside of the flight helmet against the wearer's head and neck area. The only external component was the inflation tube and port. The bladder inflated with air from the oxygen mask (via a second tube), with the increase in "G" forces.

The Thunderbirds helmets on the previous page have been retrofitted with the Combat Edge modification kit as evidenced by the inflation tube and port.

As a curious historian and collector, I contacted the Thunderbirds Aircrew Flight Equipment (AFE) shop in July of 2018 inquiring about the CE modifications. SSgt. Jasper Roberts advised in an email response that the first helmet modified was on 12 May 2018 for slot pilot #4 Nick "Khan" Krajicek[5] following his recall to the team. He advised additional helmets modified were the helmets for the pilots flying #1 - Commander / Leader, #2 - Left Wing, #5 - Lead Solo, and #6 - Opposing Solo. He added that all aircrew are projected to be fully COMBAT EDGE equipped for the 2019 show season. SSgt. Roberts did not offer an explanation for the modifications.

USAF Photo

[4] http://www.salimbeti.com/aviation/helmets7.htm

[5] Filled the vacancy left by Maj. Stephen Del Bagno.

In mid-October 2018, the USAF Thunderbirds released their accident report findings on the crash that took the life of Maj. Stephen "Cajun" Del Bagno in the spring of the same year. According to the *Air Force Times,* Maj. Del Bagno temporarily lost consciousness during a high G-force maneuver and was incapacitated right before the fatal crash in April.

Maj. Del Bagno, along with five other elite Thunderbird pilots, was practicing an aerial maneuver called the High Bomb Burst Rejoin, and flew inverted for about 22 seconds at about 5,500 to 5,700 feet above ground level, where he experienced up to negative two G-forces, the report said.

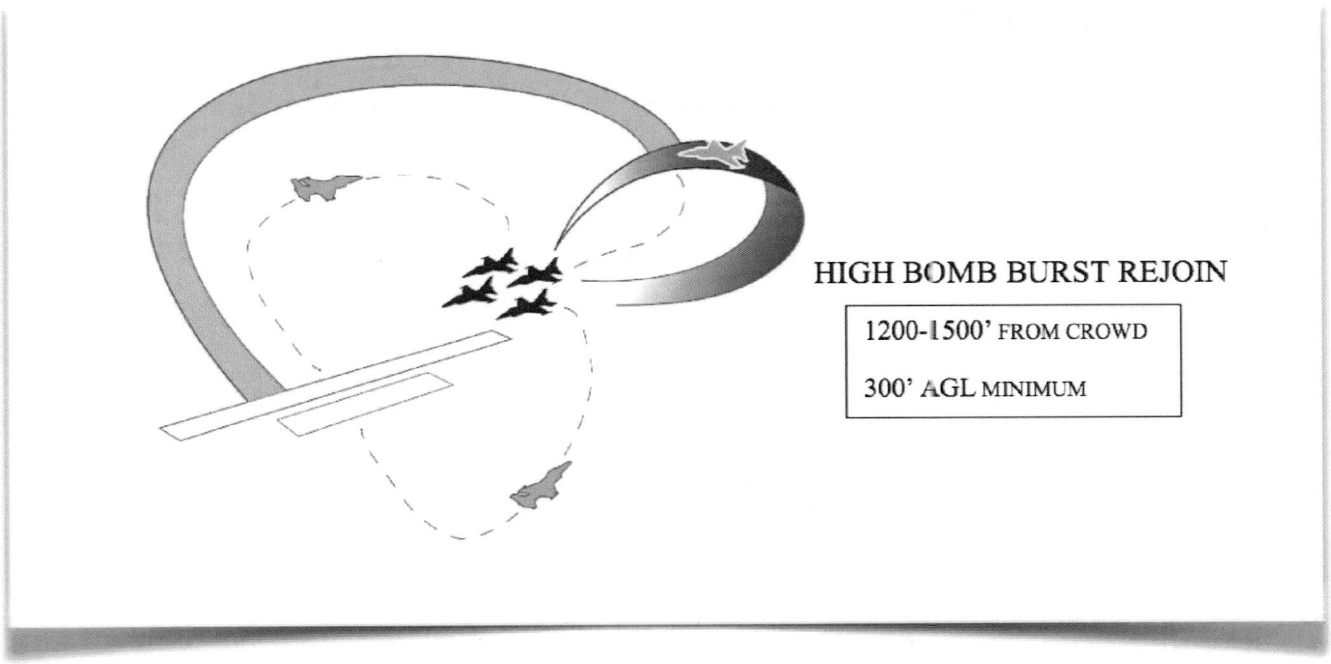

Maj. Del Bagno then started a descending half-loop maneuver called the Split-S, reaching a maximum of 8.56 Gs after five seconds, the report said. The extreme G-forces caused him to lose consciousness and be absolutely incapacitated for the next five seconds.

About a second before hitting the ground, the report said he recovered somewhat and started trying to recover his F-16CM, but it was too late. He did not attempt to eject.

The aircraft impacted the ground at 419 knots, 57 degrees nose low, 89 degree left bank at a descent rate of 39,750 feet per minute.

The accident investigation board found that the "push-pull effect" of quickly going from -2.06 G's — or twice the equivalent of standing on one's head — to 8.56 G's dramatically lessened his tolerance to G-forces, and substantially contributed to the crash. Those conditions also lessened the effectiveness of his anti-G force straining maneuver.

The USAF Aircraft Investigation Board, led by Brig. General Case A. Cunningham, states in the report that Maj. Del Bagno was not at fault for the crash. Gen. Cunningham's final words confirmed the pilot simply could not transition his body during this push-pull effect from a "diminished tolerance to positive G's induced by the physiology of exposure to negative G's".

> **"Major Del Bagno was not at fault for the crash"**
>
> Brig. Gen. Case A. Cunningham

Upon reading the accident report findings, it became abundantly clear that the Thunderbirds' decision to install the Combat Edge modifications on their flight helmets was to improve tolerance to high-G maneuvers and prevention of G-induced loss of consciousness (GLOC).

Most of us will remember him as Major Del Bagno, USAF Thunderbirds, #4 Slot Pilot. His teammates and friends will remember him as:

"CAJUN" – "BIG CAT" – "GOOD DUDE"

Rest in Peace, Sir.

As a tribute to their fallen brother, the Team had custom made pen pocket patches to be worn on the left sleeve of their work flight suits. These patches are very closely held, cherished, and not distributed outside of the immediate Thunderbird family.

Thunderbirds
4 Stephen DelBagno
Slot

8
The "Red Helmet"

As a newly selected USAF Thunderbird, pilots are required to assimilate into the team through a strict process of intense flight instruction and a right of passage before they become full fledged members of this elite team. Each of the six pilots that make up the "Delta" formation are already highly qualified fighter pilots, but must endure a comprehensive application, interview and selection process. The competition is stiff and the openings are few. In a typical year, there are hundreds of applicants which are narrowed down to about 12 semi-finalists that are further reduced by half for perhaps 3 open flying positions.

The minimum requirements (2019 - 2020 show seasons) for USAF Thunderbirds pilots are as follows:[1]

THUNDERBIRD #1 — COMMANDER / LEADER

- FLIES THE NUMBER ONE JET LEADING ALL USAF THUNDERBIRDS AIR DEMONSTRATIONS AND COMMANDS 130-PERSON SQUADRON.

  APPLICANTS MUST:

- BE A LIEUTENANT COLONEL OR LIEUTENANT COLONEL-SELECT.

- BE QUALIFIED FOR AVIATION SERVICE AND HAVE A MINIMUM OF 1000 HOURS OF FIRST PILOT / INSTRUCTOR PILOT FLYING TIME IN JET FIGHTER OR TRAINER AIRCRAFT.

---

1 https://www.ang.af.mil/Portals/77/documents/careers/MPA%20USAFE-2019-2020%20Thunderbirds%20Pilot%20Hiring%20Message.pdf?ver=2018-01-19-150806-870

- BE CURRENTLY OR PREVIOUSLY QUALIFIED IN FIGHTER AIRCRAFT.

THUNDERBIRD DEMONSTRATION PILOT (#2 - #6)

- PERFORMS AERIAL DEMONSTRATION MANEUVERS TO EXHIBIT THE CAPABILITIES OF HIGH PERFORMANCE AIRCRAFT AND SUPPORTS AIR FORCE RECRUITING AND RETENTION PROGRAMS.

APPLICANTS MUST:

- HAVE A TAFCSD[2] OF 1 DEC 2007 OR MORE RECENT.

- BE QUALIFIED FOR AVIATION SERVICE AND HAVE A MINIMUM OF 750 HOURS[3] OF FIRST PILOT/INSTRUCTOR PILOT FLYING TIME IN JET FIGHTER OR TRAINER AIRCRAFT.

- BE CURRENTLY OR PREVIOUSLY QUALIFIED IN FIGHTER AIRCRAFT.

All Thunderbird applicants — even those not selected to the team — are among the best of the best the United States Air Force has to offer. Some have combat experience; others are seasoned instructors. Despite this rarified talent pool, every new pilot joining the Team will be paired with an instructor pilot — a veteran Thunderbird pilot — and subjected to an intensive, physically and mentally challenging training program unlike any other they have experienced.

Following every show season, between December and February, the Thunderbirds go through "winter training" which takes place at Indian Springs, NV, about 50 miles northwest of Nellis AFB. Sometimes referred to as the "off season" this would be a misnomer, as the Thunderbirds arguably work harder during these 10 weeks than they do during the regular show season. Their schedule includes 2 to 3 practice flights per day, five days per week. According to the Thunderbirds, it takes approximately 100

[2] Total Active Federal Commissioned Service Date

[3] Although 750 hours is the guiding minimum, actual flight times are typically 1,500+

practice flights before the team has developed enough skill and confidence to begin a new show season in March.

The pilots new to the team, sometimes referred to as "Newbies" are not immediately issued their cherished Thunderbird flight helmets...they must be earned. These first year pilots use the USAF standard issue HGU-55/P (as discussed in Chapter 6) until they have proven themselves worthy to wear the Thunderbird helmet. This rite-of-passage occurs when their instructor pilot, and the veteran Team members, have reached a sufficient level of confidence to allow the "student" to fly his or her position solo.

When a new Team member has earned their "Red Helmet," it is an event, a celebration, a time honored tradition.

The following pages are personal reflections from various Thunderbirds, past and present, sharing stories about earning their "Red Helmet."

USAF Photos

The "Red Helmet"

(USAF Thunderbirds video production, 30 January 2018)

"Today is a fantastic day. It's an important milestone for the Thunderbirds, putting on the 'Red Helmet'. Becoming the slot pilot, or the left wing, or Thunderbird 1, or the right wing. Once you don that red helmet, it means you no longer have your instructor in the back seat. So, its a bit surreal. The safety net's gone. You're on your own, and it was just fantastic. The diamond's now together. We're going to continue to refine our maneuvers, getting prepped and ready for the delta, and in the next couple of weeks, we'll get the delta together, and we'll be ready for the show season."

Major Stephen "Cajun" DelBagno

Staff Sergeant Jasper Roberts
USAF Thunderbirds, Aircrew Flight Equipment
Nellis AFB, NV
(USAF Photo)

"We're going to give them their "Red Helmets" marking their pivotal point in their training, and progression throughout the post season in the diamond formation. To get their "Red Helmets" means that they've accomplished and overcome some big hurdles in training to where the rest of the team feels comfortable flying with them without the instructor pilot."

Staff Sgt. Jasper Roberts
USAF Thunderbirds
Aircrew Flight Equipment

Major William "Boar" Graeff
#2 Left Wing 2018 - 2019

Major Will Graeff is the Left Wing Pilot for the U.S. Air Force Air Demonstration Squadron, flying the No. 2 jet. He earned his commission in 2008, graduating from the University of Florida's, Air Force ROTC program. Before joining the team, Maj. Graeff was a T-38C instructor for Introduction to Fighter Fundamentals. Prior to his time in the T-38, Maj. Graeff was an F-15E pilot and Chief of Advanced Programs for the 492d Fighter Squadron, RAF Lakenheath, England. He has logged more than 1,400 flight hours as a USAF pilot, with over 200 hours of combat experience.

I had the pleasure of meeting with and interviewing Maj. Will Graeff in the Thunderbird VIP Room at Nellis AFB.

Maj. Graeff recalls that his first introduction to flying was when he was 10 or 11 years of age and an uncle invited him for a ride in a 4-seat Cessna, single engine airplane. He said, "I jumped at the opportunity and it was amazing!" Maj. Graeff continued, "It was a really awesome experience to sit in the front seat. My uncle actually gave me the controls to fly a bit! It was a really cool feeling…a really cool sensation and something I really cherished."

This experience influenced Maj. Graeff to pursue a career as a pilot in the United States Air Force. But, not before some soul searching.

Maj. Graeff explains, "My path into the Air Force is a little different from what you would normally find, especially with pilots. I went to college without really thinking about the military. I was going there for medical school…going to be a doctor. It's what I studied in high school. I went through a magnate program that dealt with medical aspects. But, after the first year, I realized that medicine was not something I saw for my future. And, something else I thought I was passionate about going into college was something along the lines of the F.B.I., C.I.A., some type of law enforcement career. However, I had no background or skill sets to help me for that application process. So, just kind of on a whim, I thought, well, a military background would be great in order to apply for these agencies. And, I chose the Air Force because of that scenario of flying with my uncle."

"I love flying, but again, even though that's what drew me to the Air Force, I still did not have the goal to be a pilot at that point. I was going to join the military, get some experience, then apply to the F.B.I. or C.I.A. later."

Prior to applying for a position with the Thunderbirds, Maj. Graeff was a T-38 instructor pilot at Sheppard AFB in Wichita Falls, TX, and described the application, interview and selection process for the Thunderbirds as, "Pretty in-depth; pretty detailed. Sometime in December, the hiring message will get sent out and that will get broadcast via email and the Air Force Web site."

Maj. Graeff continues, "It takes a solid month, month and a half to put the application together, which involves all of your performance reports, letters of recommendation, not only from your supervisors, but from any other people who may have influenced you, or you think may be able to put in a good word for you. In addition, a military resume, your flying history report, and all of your check rides are required to be included in the application."

"The most important thing you put together is your personal letter on why you want to join the Thunderbirds."

The packages are sent to the Thunderbirds, where the officers sort through the applicants taking note of strengths and weaknesses. Maj. Graeff said the Team is not only looking for talented pilots, but those with the right personality to work closely as a team and interact with the public, a very important part of the selection process.

"The most important thing...is your personal letter on why you want to be a Thunderbird."

Maj. Will "Boar" Graeff

The next step is to invite 12 to 14 people out for interviews. Selected candidates are flown out to an air show site on Thursday, and spend the weekend with the Thunderbirds. Maj. Graeff explains, "Not only is this a chance for the current Team to see the applicants, and see how they interact, not only with the Team, but also with the kids, as well as providing the applicants a chance to see what the Team does. It is long days, it takes a lot of energy to be able to interact constantly with the crowd. It's a chance for them to see if this is something they want to do, and a chance for us to see if they are good Thunderbird material."

Following an evaluation period, the initial candidates are cut in half and invited to return for final interviews. Each candidate will be subjected to interview questions from all 12 Thunderbird officers to determine "How they act under pressure and to see where their mental process is and their motivation for joining the Team."

When Maj. Graeff was selected for the Team, he received a phone call from the Thunderbird officers "Hooting, hollering, and cheering! It definitely gets your heart racing! It's definitely a moment of elation...something you really worked hard for and strived hard to get to...its just that culminating moment that takes your breath away!"

I asked Maj. Graeff to share his thoughts about assimilating into the Thunderbirds. He — and other pilots — were coming to the Team as experienced fighter pilots however, they are entering into a completely different flying dynamic.

Agreeing, Maj. Graeff said, "Yes, it can be challenging at times, but it's also something, with our experience level and the people we try to hire to have that humility to know, 'Yes, I'm an experienced fighter pilot, but this is a totally different flying experience' that you have no real knowledge or expertise. You need to take a step back and be on that student side of the table."

"The Team does a great job of bringing you in slowly, and warming you up to that experience. It's not so much 'I'm the instructor and you're the student' relationship, but more of a team approach. It's about how to work as a team, and how to trust each other. That team camaraderie — really family camaraderie — to be able to trust the people that are flying next to you is crucial. The Team does a very good job of building those relationships, setting the Team up for success."

"...that red helmet symbolizes more than just an event in my life...it shows that I am part of something bigger..."

Maj. Will "Boar" Graeff

On earning his Red Helmet, Maj. Graeff said, "It's a great day." "It's a culminating event for essentially a year, maybe even slightly over a year, by the time you decide to apply, went through the entire interview process and up to that point had, maybe 50 upgrade sorties, so it's a very challenging syllabus, a very challenging environment that we fly in. To be able to get through all of the steps of competing with other people to make the Team, getting through those sorties, and then to finally have that red helmet symbolizes more than just an event in my life, right, it shows that I am part of something bigger, if you will. To be able to follow in the footsteps of the legends that flew before me, starting with Buck Pattillo...to be able to follow in his footsteps and continue that mission to recruit, retain and inspire is something that I will definitely treasure for the rest of my life to have that opportunity. I think that seeing that red helmet with your name on it, and your number is something that brings it all together and really solidifies that I'm part of something bigger here, and puts on a whole different responsibility as well on your shoulders. It's not just me out there anymore. I'm representing the 660,000 people in the Air Force and that helmet symbolizes that I have to continue to do my best every single day to hopefully continue the tradition and legacy of this Team and continue to inspire the next generation.

Major Michelle "Mace" Curran
#6 Opposing Solo 2019
#5 Lead Solo 2020

Major Michelle Curran is the Opposing Solo Pilot for the U.S. Air Force Air Demonstration Squadron, flying the No. 6 jet. She earned her commission in 2009 as a graduate of the University of St. Thomas' Air Force ROTC program. Before joining the Thunderbirds, Maj. Curran served as an F-16 instructor pilot at NAS JRB Fort Worth, TX. She has logged more than 1,200 flight hours as an Air Force pilot, with more than 160 combat hours.

I had the honor and privilege of interviewing Major Curran at Nellis AFB in July 2019.

Maj. Curran says her first introduction to flying was when she was in high school. She describes growing up from humble beginnings, so when she asked her parents for flying lessons, she was disappointed, but not surprised that this was not economically feasible.

While looking for ways to pay for college, her father suggested she consider the Air Force ROTC[4] program. Maj. Curran said, "At first, I actually resisted that. I don't necessarily want to do that. I want to be a normal college student, and I want to have that 4-year college experience." However, after some consideration, Maj. Curran said, "I ended up applying and got an Air Force ROTC scholarship."

> **"...I wanted something fast, maneuverable and with an afterburner..."**
>
> Maj. Michelle "Mace" Curran

At this early juncture, Maj. Curran was still undecided whether she wanted to be a pilot, or whether she wanted to pursue a criminal justice major. As she began to meet people in both career fields, she started leaning more towards flying. As she pursued this path, and considered the available aircraft options, Maj. Curran shared that, "I was a thrill seeker growing up, I liked roller coasters, I liked going upside down, so I was like, I want to fly fighters! And, I wanted something fast, maneuverable and with an afterburner, and that's why I put the F-16 first on my list."

Maj. Curran achieved her goal of flying an F-16. Prior to applying for the Thunderbirds, she was an F-16 instructor pilot with the 301st Fighter Wing, stationed at Naval Air Station (NAS) Joint Reserve Base (JRB) Ft. Worth.

"It wasn't even something that I had 'for sure' wanted to do. I had a couple of friends from my first assignment in Japan who were on the Team so, actually the prior #4, 'Cajun' and then, our current #5, 'Spider' were both at Misawa[5] with me."

4 Reserve Officer Training Corps

5 Misawa Air Base, Japan

"So, I reached out to the them and said, 'What's the Team like'...because, I think for the 'grey world' — the combat side of the house — we don't exactly know what the Thunderbirds are like." From pilots who were on the Team, and transferred back to a grey jet, Maj. Curran says, "There are a lot of misconceptions about how difficult, or not difficult, the flying is...what all goes on outside of the actual demo...you just really don't know a lot about it until you talk to someone one-on-one." At the end of the day, the feedback was positive, and Maj. Curran made the decision to apply.

The application process is pretty lengthy, and includes "...all of your check rides you've taken in your career, your officer performance reports, its a photo, a letter on why you want to be on the Team, letters of recommendation...I believe it was 43 pages!"

Maj. Curran was selected as one of 12 candidates for semi-finalist interviews. All 12 were vying for 3 coveted openings. The semi-finalists were invited to join the Team on the road to provide them with some insight into what the Thunderbirds experience during an air show weekend. Maj. Curran said they were in green flight suits[6], and "It was cool to meet the other applicants. I knew a few of them from previous assignments in the Air Force, but got to know all of them very well. It also gives you a little perspective on who you may be potentially working with on the Team."

"The dreaded 12-man interview...Definitely the most stressful of the interviews. It was a cool experience."

Maj. Michelle "Mace" Curran

Following this initial interview process, Maj. Curran received a call that she had been selected as one of the six Thunderbird finalists. Maj. Curran added, "The semi-finalist interview was with the the 57th Wing Commander, a one-star General, and another interview with the current Team Commander / Leader and the Deputy of Operations."

6 Standard USAF issued CWU-27/P flight suit.

The stakes were higher in the finalist interviews...six pilots for 3 openings...a fifty-fifty shot. The candidates interviewed a second time with the 57th Wing Commander as well as Warfare Center Commander — a two-star General. And then, "The dreaded 12-man interview, where you sit at one end of the table...all 12 current officers on the Team are there, and they all ask you questions. Definitely the most stressful of the interviews. It was a cool experience."

Maj. Curran said she felt "pretty good" about her chances of becoming a Thunderbird, however, she added, "All 6 people that were brought back would have been great Thunderbirds, and if you are at that point in the interview process, there are no 'bad apples' at that point — everyone deserves to be there." "I definitely wasn't sure that I was going to get a phone call that I made the Team by any means, but I thought it went pretty well."

"I definitely wasn't sure that I was going to get a phone call that I made the Team by any means, but I thought it went pretty well.

Maj. Michelle "Mace" Curran

Maj. Curran — and the other five finalists — return to their respective squadrons...and wait for THE phone call.

Maj. Curran said that she had her follow-on assignment for Holloman AFB at that point, and "My report-no-later-than date was rapidly approaching. I was at the point where I should have had orders. I should have been planning my household goods to get picked up and moved, and I hadn't done any of that, because I was just waiting. I was getting emails from my sponsor — my squadron commander there — asking, 'When are you showing up...what's your timeline?' and I'm just trying to do 'chaff'[7] as much as possible, just until I know what's happening with the Team."

The phone did eventually ring. Maj. Curren said. "The caller ID was Lauren Venturini, the current Executive Officer for the Team. I knew for sure what this was about!

[7] Chaff is a countermeasure deployed by pilots to confuse the adversary's radar.

When she answered the call, "The whole Team was on speaker phone shouting congratulations! It was awesome! I was elated, for sure! I was walking on clouds the rest of the day."

Maj. Curran continues, "And then, the whirlwind begins, because once you find out, it's a short timeline to change all my orders that had me going to Holloman, get them set up to come out here to Nellis...find a place to live. It's definitely a quicker transition than you normally have. Full flail at that point!"

As soon as she arrived at Nellis AFB, Maj. Curran and the other two new Thunderbird pilots joined the Team on the road for the final 2 or 3 shows remaining in the season. She says, "It was cool from the perspective of knowing that I was going to do it, and knowing what position I was going to have." She continued, "When you apply, you are asked which position is your preference, but, if everyone puts the same thing first, they have to figure it out. To find out that I was going to be a solo, which was my first choice, was amazing!"

When I asked Maj. Curran if she was enjoying her new job, she replied, "It's awesome! The flying is a lot more difficult than I expected. Previously flying the F-16, when you're brand new to the jet, of course, you learn the handling of the aircraft, the stick and rudder skills…it's a pretty easy jet to fly overall, and when you're getting further into your career, and you're really into the 'tactical weeds' you're just managing sensors, all your different pods, your radar, all that stuff…the actual physical flying, you just don't even think about."

"To find out that I was going to be a solo, which was my first choice, was amazing!"

Maj. Michelle "Mace" Curran

"But, on this Team, you don't have anything but the flying. So, its difficult…a lot of rudder that we almost never use…I'm doing 'full boots of rudder' either direction several times during the show…a lot of flying upside down, which we don't normally do…but, it's fun! Once you're proficient at it, it's a lot of fun, but it can be stressful at first."

I was surprised to learn that Maj. Curran's first exposure to the Thunderbirds was after she joined the Air Force. "I never saw the Thunderbirds until I was already a student [pilot] at Columbus AFB and had just gotten what we call our 'Drop Night' where we find out what airframe we're going to. Probably a few weeks prior, I found out I going to F-16s, then the Team [Thunderbirds] came to Columbus to do an air show. That's the first time I saw the Team. And, it's funny, at the time, 'Mother' Hubbard[1] was on the Team — number 8 — and, one of my friends who was going to C-17s said, 'There's a girl on the Team, there's a female pilot…this is awesome!'"

[1] Then Captain Kristin "Mother" Hubbard

"We were at the O-Club for the Team's P.R.[9] and I was just a second lieutenant, and I was too scared to talk to her because she was intimidating, which is so funny now because I've actually overlapped, or run into her throughout my career, and she's one of my really good friends now. I told her that story, and she's like, 'You're such a dork!'" "Now, here I am in a blue flight suit flying for the Team 8 or 9 years later. It reminds me that my interactions with people maybe something that's lasting for them down the road."

Maj. Curran shared some interesting insight about assimilating into the Team as a new Thunderbird pilot. The typical commitment for a numbered pilot is a two-year assignment. "In any given year, half of the pilots are rotating out every single year, so it's a pretty well oiled machine as far as the new pilots coming in, and others moving on

9 Public Relations

to whatever is next in their career. People are really welcoming, and they know what has to happen in the training season, and they also know how demanding it is and very understanding of that. They know the demand it puts on your family, on your time outside of work, and then, your body. Sometimes, we fly the Demo 10 times in a week. It's a whole other ball game on what you need to do to keep yourself healthy. Your sleep, your work-out plan, all that stuff."

"...we walk into The Perch...and all of our red helmets are lined up on the bar...I didn't even know this was happening today!"

Maj. Michelle "Mace" Curran

"Assimilating into the team [for the pilots] is very easy. You know a lot of them... they are your peers. I think getting to know everyone else on the Team, and getting them to know you, and gaining their respect for you, because you are the face of their Team, I think that takes a little longer...that probably takes the entire training season before you are accepted as, 'This is our Team for the 2019 Season.'"

"As the flying goes, they realize the job they have ahead of them to train you, and then they realize the obstacles you're going to face learning the profile over the training season."

On earning her red helmet, Maj. Curran recalls, "I didn't even realize I was getting my red helmet that day, actually, because the diamond was coming together, and they were on their last syllabus sortie to officially be done training as the diamond. As a solo, I'm running on a parallel syllabus at the same time. I'm doing most of my training just as a two-ship flight with #5." She thought, "I wonder if I'm going to get it [red helmet] when the delta comes together...the Superbowl is coming up really soon, obviously we're going to have our red helmets for that! Hopefully this happens soon, but I don't know for sure!"

"But, the diamond came together, and I was almost done with my syllabus at that point, which is about 90 flights. It's a lot that we do over one training season. And then, we walk into 'The Perch,' and the whole Team is there, and there's a cake, it's diamond day, everyone is coming together, and all of our red helmets are lined up on the bar." With excitement in her voice, Maj. Curran said, "I didn't even know this was happening today! There's a tradition where #3 takes a giant cleaver and chops up the cake and it flies everywhere!"

"That was cool, because you taxi around the airfield in a Thunderbird jet, and you use a call sign on the radio, and everyone knows who you are, so you're kind of under a microscope. That was kind of like the final piece, to fly around with the red helmet on."

"To go from flying with the grey helmet, to showing up in a show suit and the red helmet…those are the two big things…this is really happening…we are about to go do this all over the United States for the next eight months."

"Everyone is excited at this point. Morale is very high. We have reached a tipping point where we are going to focus on show season and actually showcase what we've been working on for the last 4 or 5 months."

Lt. Col. Caroline "Blaze" Jensen

Lt. Col. (then Major) Caroline "Blaze" Jensen became the third female demonstration pilot, and first female reservist selected to the USAF Thunderbirds. She flew the #3 Right Wing position for the 2012 to 2014 show seasons. She has accumulated over 3,500 flight hours, and more than 200 combat hours in the F-16 "Fighting Falcon" over Iraq. Impressed? Add "Mother" to her resume!

In July of 2019, Col. Jensen graciously granted me an in-person interview at the Thunderbird VIP Room. Her story is inspiring.

Col. Jensen is currently the Deputy Commander of the 926th Operations Group at Nellis AFB, and says her interest in aviation was first inspired by her father, who was a Marine H-34 helicopter pilot in Vietnam.

She recalls her first distinct memory of wanting to fly at about the age of six. Her parents were watching a movie, and remembers a biplane flying among big, puffy clouds. Col. Jensen said, "I was really enamored with the idea of cloud chasing. That's what really hooked me."

She was influenced by several family members who also served their country in the armed services; however, she said her primary influence to join the U.S. Air Force was "To serve…and to be a part of something bigger than myself."

Photo credit: SSgt. Larry Reid, Jr.

Col. Jensen shared that her first exposure to the Thunderbirds was when she was 12 or 13 years of age and her father took her to an air show in Eau Claire, WI. Upon watching the Thunderbirds perform, she told her father, "I want to be a fighter pilot!"

Fortunately, by the time she was a Senior in high school, the combat exclusion for women had been lifted. She was later accepted, and graduated from the U. S. Air Force Academy, clearing the way for her to achieve her goal to serve as a fighter pilot for the U.S. Air Force, but not without a fight. Col. Jensen admitted that her first year at the Academy was academically challenging. Following freshman finals, the Thunderbirds arrived for a scheduled performance, and one of the jets impressed her as it screamed low-level between the chapel and the front range. This was an inspirational moment, and just the motivation she needed to refocus. Little did she know at that time the significance of this event.

Following a combat tour in Iraq where Lt. Col. Jensen logged more than 200 combat hours in the F-16 "Fighting Falcon," she became pregnant with her son, so transitioned to a reserve unit at Sheppard AFB, TX flying T-38s, which she refers to as "One of my first loves."

Col. Jensen said, "I'm a very 'goal oriented' person" and decided she wanted to be a fighter pilot prior to women being accepted in fighter cockpits. Having achieved her first goal, her next goal came when she was reading the U. S. Air Force Reserve publication, *Citizen Airman Magazine*, and came across an article about the first USAF Reservist selected as a Thunderbird pilot! She exclaimed, "That's what I want to do!" And, that's what she did!

When Col. Jensen submitted her application package to the Thunderbirds, there had only been two female demonstration pilots to serve on the Team. She impressed the Team, and was one of 12 candidates selected for a first round of interviews, which involves traveling to, and "shadowing" the pilots at a scheduled air show. The candidates arrived on Wednesday with the advance pilot, assisted with Team "arrivals" on Thursday followed by media interviews, a school visit on Friday morning, a practice show Friday afternoon, a public relations event Friday evening, and, air shows on Saturday and Sunday. Add to this, interacting with the media and air show spectators on the autograph line! This was the candidates first taste of what it was like to be a Thunderbird.

USAF Academy, Cadet Chapel (USAF Photo)

Following this "orientation," the candidates returned to their respective squadrons, and about two weeks later, the 12 finalists were reduced to 6. Col. Jensen's commander called to notify her that she made the cut! The remaining 6 were invited back for a second air show "experience" after which, the finalists are subjected to the "12 man interview" back at Nellis AFB.

One at a time, the candidates report to the Thunderbird VIP Room in service dress for formal interviews. The "numbered" Thunderbird officers are seated at the conference table in their respective places. The candidate takes a seat at the opposite end of the table from the "Boss" and field questions from each of the officers. Unnerving, to say the least.

"The flying we do [Thunderbirds] is like nothing else we do in the Air Force."

Lt. Col. Caroline "Blaze" Jensen

The candidates once again return to their squadrons, and await "the phone call." Col. Jensen recalled that she was vacationing at Disneyland when her commander called to let her know she had been selected to the Team. She received another call shortly thereafter from the Thunderbirds, with everyone cheering, "Congratulations, you made the Team. We're so happy to have you!" Col. Jensen said she felt like her feet weren't touching the ground, but the responsibility of tending to her 2 1/2 year old son soon brought her back to reality…for the moment!

When we discussed assimilation into the Team, Col. Jensen said that "The flying we do [Thunderbirds] is like nothing else we do in the Air Force. There are a lot of really hairy moments where you're consciously trying to fly good formation position but your subconscious is like, 'Fight or Flight' and it's time to get out of here! You just keep going at it, and it all works out, eventually. It's a really amazing process. It's very tough, mentally and physically, and probably one of the most exhausting things you can do in a jet."

Col. Jensen shared that, "The training season is a building block process. You start out with 2 planes, eventually adding more planes, all the while reducing the altitude at which you perform the maneuvers."

"During that process, we are wearing the grey helmets.[10] We did get fitted for our red helmets early on, but we had to wait. We progressed from 2 aircraft, to 3 aircraft, to 4 aircraft. The Diamond comes together when the upcoming season's pilots flew without the instructors in the back of the 'D' model, or chasing the formation. Then you progress to getting the Delta [formation] together."

"It's a really amazing process. It's very tough, mentally and physically, and probably one of the most exhausting things you can do in a jet."

Lt. Col. Caroline "Blaze" Jensen

"I'm pretty sure I got my Red Helmet when the Diamond formation came together. The cool thing that I remember is that we never touch our life support gear. We don't do a 'walk-around' the airplane as a pilot. We have this awesome crew of life support, egress people, and crew chiefs who take care of everything."

"The first time I see my Red Helmet is when I march to the jet and put it on — it was a really good feeling!" This "right of passage" was subsequently celebrated in "The Perch" with a cake to commemorate the occasion, a practice that continues today.

I asked Col. Jensen what the "Red Helmet" meant to her, and she said, "Oh my gosh, there's a lot of history. I feel really honored that I get to wear the same helmet as Hank Canterbury[11] who was the first Air Force Academy graduate to fly with the Thunderbirds. I had a great mentor, Matt Byrd[12], who just retired last year and had been Thunderbird #3, 20 years before me."

---

[10] USAF issued HGU-55/P

[11] Henry Canterbury, #4 Slot Pilot, 1965-1966.

[12] Matthew Byrd, #3 Right Wing, 1993-1994.

"I was like one of the first thirty-something pilots to fly in the right wing position. Buck and Bill Pattillo were the first two pilots to fly the left and right wing. It means a lot to me on a lot of different levels."

When asked how she juggled being a Thunderbird and a Mom, she said, "It was very challenging!" Col. Jensen credits a supportive family during this time in her career. In typical Thunderbird fashion, Col. Jensen said "A lot of people were very interested in me as a woman and my story and would ask me all of these questions. I feel it's the job of a Thunderbird to take that light that's shined on them and to display all the other people who are not only on the Team, but also in the Air Force. Everyone is working very hard everyday to make the mission happen."

Col. Jensen reminisced about her last show as a Thunderbird[13], with her son Finn in attendance along with her father. A local reporter asked Finn "So, what is your favorite part of the air show?" Finn replied "When my mom flies." With slight disbelief, the reporter said "Oh, really, what does your mom fly?" Finn said "She flies one of those" as he pointed to one of the Thunderbirds parked in the background. When the story aired, they cut to a Thunderbird pass and said "For the record, his mom flies one of these." A very proud moment, indeed.

Col. Jensen concluded, "People would ask me all these questions about me and my story and what its like to be a woman, one reporter wouldn't stop asking about it, and I said, you know what, women have been flying combat for over 20 years, we've had women on the Team for a long time now. The thing that I am most proud of is I'm the first pilot to be a Mom on the Team."

The next day, Col. Jensen showed up for work and had several newspapers in her briefing spot, one with the headline: **"Thundering Momma!"**

---

13 Nellis AFB Aviation Nation 2014

CALIFORNIA INTERNATIONAL AIRSHOW SALINAS

Thundering Momma

PROVIDED COURTESY CAROLINE JENSEN

U.S. Air Force Reserve Major Caroline Jensen holds her 5-year-old son, Finn, near her F-16 Fighting Falcon. Jensen flies the No. 3 right wing slot for the Air Force's Thunderbirds Air Demonstration Squadron.

Pilot's life as mom, Thunderbird

By Jeff Mitchell
jemitchell@thecalifornian.com

U.S. Air Force Reserve Maj. Caroline Jensen hopes someday that the fact she is a fighter pilot and a mother of a 5-year-old son someday won't be such a surprise to people.

After all, she reasons, there's lot of fathers serving in all branches of the American military today.

But add in that she's a member of the U.S. Air Force's elite Thunderbirds Air Demonstration Squadron and, like it or not, she tends to stand out a bit.

But if it's a unique situation then it's a situation she says she relishes.

"I love doing it all and I am so grateful to the Air Force for all the opportunities that have been offered to me," the 38-year-old Wisconsin native says. "I can't tell you how proud I am to be a Thunderbird and to fly and work with this squadron."

Jensen is the first female Air Force Reserve Officer to fly for the storied military aerobatic team, as well as the first mom. She also is currently the only female officer in the 120-member unit.

Jensen, pilot of the No. 3 aircraft, and her fellow Thunderbirds said they look forward to appearing at the California International Airshow Salinas next month.

She and her teammates last

See PILOT, Page 3A

IF YOU GO:

The California International Airshow Salinas will be held Sept. 27-28 at the Salinas Municipal Airport. For tickets and other information, go to the show's website at *http://salinasairshow.com/* or call 831-754-1983. Gates open each day at 9 a.m. and flying begins at noon.

BIOGRAPHY:

Name: Caroline Jensen

Thunderbird call sign: 3 - Right Wing

Informal call sign: Blaze

Rank: Major, U.S. Air Force Reserve

Age: 38

Born and raised: River Falls, Wisconsin

Married to husband: Shane

Children: Finn, 5 years old

Hometown: Las Vegas, Nevada

Hobbies: Swimming, running, lifting weights.

PROVIDED COURTESY CAROLINE JENSEN

While his mom flies her F-16 Fighting Falcon in formation at more than 500 mph, Finn Jensen enjoys a day at the beach in a recent family photo. Jensen's mother, Caroline, flies the No. 3 right wing slot for the Air Force's Thunderbirds Air Demonstration Squadron.

The Salinas Weekend Californian, August 23-24, 2014

Major Trevor "Dozen" Aldridge
#2 Left Wing - 2020

In November of 2019, I had the distinct privilege of attending the USAF Thunderbirds Alumni Association (TBAA) Reunion at Nellis AFB. This biennial event celebrates all Thunderbirds — past and present — in grand style over several glorious days.

The festivities began on Thursday evening, where attendees got acquainted at a cocktail social, and continued into the wee hours in a dedicated hospitality suite at the host hotel. The following day, we were treated to a USAF Thunderbirds performance and open house at Nellis AFB, an event I will not soon forget!

Following the air show, the Thunderbird pilots — incoming and outgoing — were mingling with guests in their hangar. This is where I met Major Trevor "Dozen" Aldridge, the new #2 Left Wing pilot for the 2020 season.

(USAF Photo.)

My first impression was Maj. Aldridge's youthful appearance — too young, by my estimation — to have become an experienced fighter pilot, achieved the grade of O-4, and acquired the demanding credentials necessary to secure a coveted position on the Team. Perhaps I was in denial that I have children who have celebrated more birthdays! I quickly warmed up to Maj. Aldridge's infectious personality, zeal for aviation and passion for the Thunderbirds — qualities that epitomize the Team's role and mission.

Later that evening, the TBAA hosted another cocktail reception, followed by a "Dining Out" dinner that was attended by current and former Team members, retired general officers, family and friends. Of course, the hospitality suite was open "until" to allow the continued camaraderie among this unique fraternity.

Not one to miss an opportunity, or an open bar, I wandered over to the hospitality suite. When I arrived on the appointed floor in the hotel, I began walking down a very long hallway. I was guided by the jovial sounds of laughter and spirited conversation emanating from an open door. Along the way, I began to recognize some familiar faces I had met earlier. I was in the right place.

Major Trevor "Dozen" Aldridge. USAF Photos.

I surveyed the room, navigated my way through a sea of people to the bar, and secured an adult beverage. As I made my way back towards the door, I once again ran into Maj. Aldridge. He remembered our earlier conversation at the hangar, and we picked up where we left off. The conversation soon turned to my affiliation with the Thunderbirds. I explained that I was a guest of the TBAA as a result of my research on the history, development and evolution of the USAF Thunderbirds flight helmets, and upcoming publication of my book, *The Red Helmet.*

Maj. Aldridge listened intently, and seemed genuinely intrigued, as I shamelessly self-promoted my latest pet project. Imagine my surprise when he asked if he could solicit my professional opinion on an idea he had regarding a change to the Thunderbird helmet. Needless to say, he had my undivided attention. He was quick to qualify what he was about to reveal by stating that he understands and respects tradition, adding that his intention was to introduce a tasteful modification.

He asked if I was familiar with the gold mirrored visors used by their sister flight demonstration team, the Blue Angels. Of course. These iconic visors have been in use since the early 1970s. Maj. Aldridge asked, "What do you think about the Thunderbirds using a silver mirrored visor?" My immediate reaction was, "Brilliant!" In addition to the aesthetic improvement, Maj. Aldridge imagined that the reflection would create some incredible photo opportunities. I was immediately sold on the idea and eagerly inquired about the next steps.

Maj. Aldridge advised that he located a supplier for the silver mirrored visors and placed an order for one, intended for "show-and-tell" in an attempt to "sell" the other 11 officers.

The supplier expected to ship the visor in January of 2020. Meanwhile, Maj. Aldridge used his time to research any applicable regulations and Air Force Instructions (AFIs) that may be applicable to his proposed change. He polled Thunderbird officers #3 through #12 and all were overwhelmingly in favor of, and supportive of the silver visors. The "Boss" would be the final authority, making sure all the details were in order before the "big reveal". At the end of the day, Boss would cast the final vote.

I shared with Maj. Aldridge that if approved, aside from the safety improvements made in 2018 (Combat Edge modifications), this would be the first aesthetic change to the Team's flight helmets in a very long time. If I had a vote in the matter, it would be a resounding "YES!"

It is interesting to note that there is a commercially available USAF Thunderbirds, half-scale, "mini flight helmet" produced by *FOX-2 Flight Gear.* Sold as a collector's item, or for obtaining Team autographs, the manufacturer has exercised some artistic license and produced these helmets with a silver mirrored visor. I think this is a winner, for sure!

The photo on the following page (top left) is an example of a Thunderbirds "mini flight helmet offered by *FOX-2 Flight Gear.*

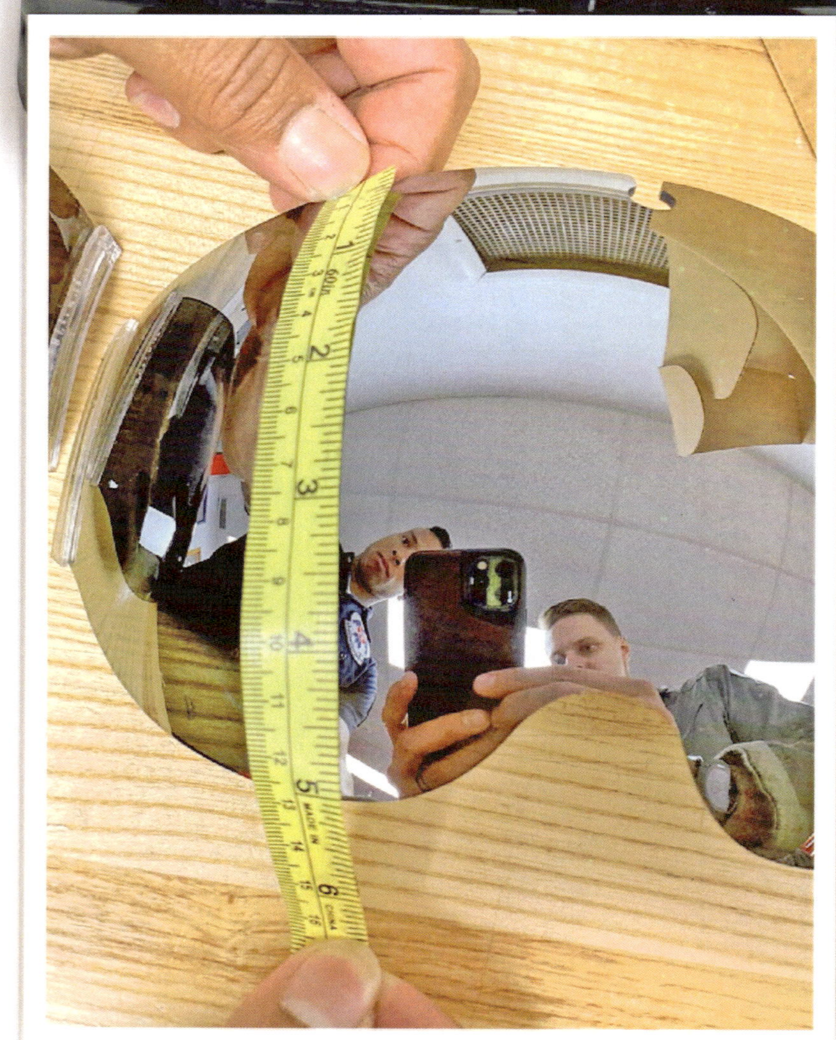

The reflective quality of the visor is stunning, and complements the chrome oxygen mask bayonet receivers — a winning combination in my book!

The bottom photo is the first silver mirrored visor delivered to Maj. Aldridge for test and evaluation. The reflective qualities of this visor clearly shows Maj. Aldridge and AFE Specialist SSgt. Kyle Boddie conducting a critical examination prior to fitting it to a Thunderbirds flight helmet.

Unfortunately, this example did not meet the exacting military specifications required by Thunderbirds AFE.

Back to the drawing board.

Given the demanding schedule of the Thunderbirds, I offered to assist Maj. Aldridge in his search for an appropriate company, preferably U.S.-based, to provide this service. I was also told that their preference would be to send their government-issued, genuine GENTEX, Mil-Spec visors to the successful vendor, rather than rely on on the supplier to provide the visors and coating. I agreed to keep the identity of the end-user — the Thunderbirds — confidential until further notice.

My first order of business was to contact James Wegge, Director of Gibson & Barnes for some professional direction. I learned that the Blue Angels send their visors to Oakley Sunglasses in Foothill Ranch, CA for gold mirrored coating. Mr. Wegge provided me with a point of contact — military sales specialist — and felt certain Oakley could do the silver mirrored coating as well.

Despite my best efforts, I failed to make the connection, so Maj. Aldridge agreed to give it the old college try, and called a fellow Texas A&M University "Aggie" with the Blue Angels, Frank Zastoupil. It turns out the Blue Angels contract with Oakley has since moved to Eyeking LLC, a subsidiary of Under Armour Eyeware. Maj. Aldridge called Eyeking and – not surprising – they agreed to provide the Thunderbirds with silver mirrored visors! Maj. Aldridge delivered his proposal to Lt. Col. John Caldwell – better known as "Boss" – and received the "green light" to proceed!

Maj. Aldridge said that they will be placing one silver mirrored visor into service for additional testing by "Boss" while awaiting approval from the Wing Commander. Once final approval has been granted, it may take a few months to requisition new visors from Gibson & Barnes, and send them to Eyeking for coating. The goal is to have all of the visors upgraded no later than mid-year of the 2020 show season.

Well done, Sir! I applaud your vision (pun intended) and will be looking forward to seeing the entire Team in their new, "tricked out" **Red Helmets!**

Major Trevor "Dozen" Aldridge earned his "Red Helmet" in January 2020 following the successful assimilation of the "Diamond" formation. This marks a huge milestone for the "noobs" as they have earned the trust and respect of their instructors and veteran Team mates and have become full-fledged Thunderbirds!

9
Call Signs

There is a long-standing tradition in the aviation community for pilots to be given nicknames, or "Call Signs" as they are known in the military vernacular. With so many references to call signs in the previous chapters, the time has come to take a closer look at these curious, and sometimes embarrassing monikers.

According to an article on the Air Force website by Second Lieutenant William Powell, the first call signs were probably used by ground controllers to communicate with pilots during World War I. "It was faster and easier to call a pilot by his nickname, and it would have confused the enemy in case they were listening," says Yvonne Kincaid, an Air Force historian. By 1930, it was common for pilots to have a unique call sign.

Flight helmet photo courtesy of Pete Mitchell, California.

Call signs almost always come from another member of the squadron or a friend, and is typically "awarded" as a play on the pilot's name, or previous exploit in the pilot's past. It is also considered bad form to give yourself a call sign. It is not uncommon for the call sign to be derogatory, and the more the pilot complains, the more likely it is to stick. Usually, once a pilot flies with a call sign in combat, they get to keep it for their career.

Most of us are quite familiar with the 1986 box office hit, *TOP GUN*, and the prolific use of call signs throughout the movie. Who doesn't remember LT Pete "Maverick" Mitchell, LT Tom "Iceman" Kazansky and CDR Mike "Viper" Metcalf, just to name a few?

The Thunderbirds are no exception to this practice and have been using call signs on their flight helmets since the 1950s. When I spoke with the Thunderbirds Life Support Shop[1] in the mid-1990s, I was told at that time the pilots had a choice when it came to personalizing their helmet visors: First name, or call sign. During the 1990s, there seemed to be more first names used, but currently, call signs are the more common practice, as evidenced in Chapters 6, 7 and 8.

Some pilots use the same call sign throughout their careers, while others may have used several. For example, U. S. Navy Lieutenant Kara "Hulk" Hultgreen received her call sign because of her ability to bench-press 200 pounds. Following a television appearance in which she wore significant makeup, her squadron mates rewarded her with the revised call sign "Revlon."

Traditions and ceremonies for assigning call signs vary between the U. S. Navy and the USAF, and from squadron to squadron. Some hold a "naming ceremony," usually after a new squadron member, or members — called "FNGs" for F'ing New Guy / Gal — are gathered together with their fellow squadron mates for drinks, heckling and bribes.

According to Chris Kibble[2], "Bribing by the FNGs is encouraged in preparation for the naming ceremony…to the people with the final say for your call sign. Bribes are an art, not a science. They should be respectful but not audacious. Generous but not out of hand. You're trying to get right up to the line of kissing the committee's butt without crossing over. Alcohol is always welcome, as is food. Bribes have changed over the years as squadrons have evolved. Items that were apparently mandatory for past

[1] Currently Aircrew Flight Equipment.

[2] Former USAF F-16 pilot.

generations would get you kicked out of the Air Force today. But the spirit remains the same."

Some common guidelines in the naming process involve choosing call signs that are not very flattering, and usually bring attention to something embarrassing the FNG has done. It should be "safe" enough to be shared in mixed company, but that rule is not always followed! Trust me, I know many inappropriate call signs not suitable for this book!

The USAF Thunderbirds pilot, flying the #3 (Right Wing) jet for the 2006 - 2007 show seasons was Maj. Nicole Malachowski. Her call sign — "FiFi" — while not necessarily very flattering for a woman, is typical ridicule, and generally tolerated in the fighter pilot community; however, it could be a shortened version of "Fifinella" the imaginary "gremlin" of WWII fame that became the mascot of the Women's Air Service Pilots (WASP). A third version suggests whenever she made flying mistakes, she would scold herself by repeating, "F'ing Idiot...F'ing Idiot!"

I am fascinated by call signs, and I learned the hard way that if one is not privy to how a pilot received their call sign, then asking is not appropriate unless you are in a semi-private setting — typically a bar — and are prepared to purchase the pilot a drink in exchange for the reveal. The first time I boldly asked a pilot the story behind his call sign, I was with Lieutenant Colonel Greg "Growth" Wilson (photo above), a neighbor back in the late 1990s. Col. Wilson, a

Raven Forward Air Controller[3] flew O-1 "Bird Dogs" out of Laos during the Vietnam War, and later flew F-16s and A-10s with the 706th fighter squadron at Joint Reserve Base, New Orleans. Col Wilson also served in Operation Desert Shield (2 August 1990 – 17 January 1991).

When Col. Wilson graciously presented me with an inscribed and autographed copy of the book, *The Ravens,* I unwittingly inquired, "Sir, do you mind if I ask how you got your call sign?" Without blinking an eye, he barked gruffly, "Yes I do!" Thank you sir… have a nice day…end of story!

Fast forward to January 2019. As one of 3 new pilots on the 2019 Team, Michelle "Mace" Curran, USAF Thunderbird #6, Opposing Solo, was interviewed by *Airshow News."*[4] The very first question raised was, "Call sign, and how'd you get it?"

Capt. Curran provided a text book response: "You know the rules! We can't just throw that info out all willy nilly. Traditionally, the stories behind call signs are only told over a beverage to a trusted few." Well played, Mace!

Thankfully, not all pilots hold their call signs so close to the vest. A very good friend of mine, Pat "Flaps" Falterman and I attended the same high school and have shared an interest in aviation since that same time. As a "green" private pilot, I gave Pat his first ride in a Cessna C-172 in 1979, and as we all have at one time or another, fell in love with flying. Through college ROTC, Pat earned a commission as a 2nd Lieutenant in the USAF, completed Undergraduate Pilot Training (UPT) flying T-37s and T-38s, and later became an aircraft commander on the KC-135 and T-43, as well as an instructor pilot. Pat is still in the cockpit today as a Captain of a major U. S. air carrier, based in the Southwest.

Pat is the best storyteller I know and is never without a tale to share, regardless of how humiliating or embarrassing it may be to him, or in many cases, his audience! Pat's

---

3 The Raven FACs were a covert Central Intelligence Agency operation tasked with identifying and marking ground targets for U. S. air strikes in Vietnam.

4 https://www.newsairshow.com/headlines/2019

number one rule of storytelling is, "Never allow the facts to get in the way of a good story!" Over the years, he has held firm to this rule, and his stories continue to be second to none! He reminds me of a well seasoned stand-up comedian, delivering a carefully rehearsed routine to an eager crowd. His pauses are meticulously calculated, allowing ample time for the roar of laughter and applause to subside before setting up for the next punch line. He is unfazed by the booing and jeering that inevitably arise when the story takes an unflattering turn. This guy is good...perhaps a legend by some accounts. Add alcohol to the equation and the stories become infamous.

"Never allow the facts to get in the way of a good story"

Pat "Flaps" Falterman

When I pitched the idea to Pat about sharing his personal story of earning his call sign, he predictably jumped on the opportunity! Pat's call sign — "Flaps" — was awarded during his early UPT days flying the Cessna T-37 "Tweet" that some refer to as the "6,000 Pound Dog Whistle" because of the high pitched, ear piercing sound of the engines.

Not surprising, when Pat submitted his written account of how he earned his cherished callsign, he began with a story before the story...the opening act, if you will!

Pat echoed some of the background surrounding call signs mentioned earlier in this chapter, such as how a pilot's surname is often the impetus behind this time honored tradition. Pat was in the Air Force from 1986 to 1998 and remembered two call signs from UPT at Williams AFB, AZ that quickly came to mind as examples: One student's last name was Kasberg, so he was known from that point forward as "Ghost." The second student's call sign was

"Chief" because he was of Native American descent — the name that Pat calls him to this day!

"I flew KC-135A Air Refueling aircraft in SAC and I do believe back then the use of and attachment to call signs differed from command to command and may still to this day."

Pat continues, "The use of call signs outside of the military setting, and still using old military call signs when in a civilian capacity, is usually frowned upon unless the other pilot is former military and plays along. While riding jump-seat one day, I watched the crew meet and the older captain asked the new first officer his name. The guy responded with "Maverick" and the old captain said, 'Unless that's actually on your birth certificate how about I call you what your momma named you?!' However, in an informal setting, call signs are still used basically forever and I struggle to remember some of my old friends real names at times!"

Pat digresses, "Back to my call sign and how I 'achieved' it."

"Well, after the whole preflight dance was accomplished, I attempted once again to lower the flaps and the damn lever wouldn't budge!"

Pat "Flaps" Falterman

"I was a newly soloed UPT student heading out to fly my first sortie after solo, alone and unsupervised. My briefing from my instructor pilot (IP) consisted of 'Don't do anything Dumb, Dual (certain maneuvers where supposed to only be done with an IP such as a spin for example), or Different, and if you screw up on the radios fess up. We all know it's you!'"

Pat explains, "Being from South Louisiana, and having the unique accent of the area[5], it was impossible to hide when I keyed the mike. My first IP was from Boston and he threatened to put me in the English speaking class with the student pilots from Kuwait!

[5] Cajun Country

I didn't dare tell him I was having a real hard time understanding him also; you didn't bother the IPs much back then."

"So, there I was, headed out to the flight line alone for the first time. I preflighted the T-37 as trained, got strapped in with the help of the crew chief, and cranked up the engines. However, when I attempted to lower the flaps, the handle would not budge. I tried everything and was convinced I missed some sort of gust lock or neglected to turn on the hydraulic pumps, but nothing was out of order. All I could do was shutdown and dread taking the walk of shame back to the hanger."

"As I was unstrapping from the jet, the crew chief said there was another jet down the line I could have if I still wanted to fly. I quickly agreed and was amazed he was not blaming me for the first aborted jet, and that he still had some confidence in my proficiency after my whopping one hour of solo time. Well, after the whole preflight dance was accomplished, I attempted once again to lower the flaps and the damn lever wouldn't budge! By this time I was positive I was somehow screwing the pooch, and by the look on the crew chief's face he seemed to agree!"

"This time, I completed the walk of shame to the hanger while my mind raced through preflight procedures wondering what I had missed. As I entered the scheduling room amid all the chaos of IPs and students coming and going, I caught the eye of my IP who was also a scheduler. He asked why I was still on the ground and I tried to explain what had happened. He didn't have time to hear my story and just told me to write on the scheduling board as briefly as possible why I aborted two jets as he was in a hurry to go fly a sortie himself. Needless to say I was flustered and

"This is the true story of my call sign contrary to the attempts to blame the name on my gift of gab and storytelling which may have some validity!"

Pat "Flaps" Falterman

scheduling boards are a nightmare of names, times, sortie types, maintenance status, etc., all written in some type of shorthand known only to the schedulers. I found my line on the scheduling board and wrote, 'FLAPS, FLAPS' as the reason for the day's aborted attempts. Figuring this was as succinct as I could get to explain the problem, I quickly made my exit so I could get in the books and find out what I had really done wrong."

"The next day at 'Standups' (which is an evil oral exam given every morning by the IPs), a 'Lt. Flaps' was called to standup and begin to work the morning's situation. We all looked around as we knew there was no one by that name in our group for sure. It was then that the IPs unveiled the scheduling board and where my name was supposed to be was first name Flaps, last name Flaps. In my haste and inexperience I put the malfunction explanation in the place of both my first and last name which it seemed to them to be absolutely hilarious!"

"So, my call sign from then on and to this day has been Flaps. This is the true story of my call sign contrary to the attempts to blame the name on my gift of gab and storytelling which may have some validity!"

"As an explanation of why the flap lever would not move during preflight, it turned out at least this time it wasn't my fault! A new maintenance guy performing the post day flight checks had used the wrong tension test on the flap cables and overtightened them. There were several other jets that had the same problem that morning. Needless to say this vindication did not prevent my call sign from sticking to this day!"

Pat "Flaps" Falterman, ladies and gentlemen! He will be here until Thursday! Try the veal, and don't forget to tip your waitress!

A few of my favorite Thunderbird Call Signs[6]

Gen. Lloyd **"Fig"** Newton
Capt. Jeff **"Roach"** Rochelle
Capt. Clark **"Buck"** Rogers
Maj. Shawn **"Norm"** Pederson
Maj. Brian **"Rug"** Burns
Maj. Nicole **"FiFi"** Malachowski
Maj. Samantha **"Combo"** Weeks
Maj. Kirby **"Gat"** Ensser
Maj. Richard **"Face"** Goodman
Lt. Col. Case **"Basket"** Cunningham
Maj. John **"Nuke"** Gallemore
Capt. Kristin **"Mother"** Hubbard
Maj. Joshua **"Tater"** Boudreaux
Maj. Jason **"Jaws"** Curtis
Capt. Michelle **"Mace"** Curran

I do have one more favorite call sign, and this book would not be complete without including Maj. Stephen "Cajun" Del Bagno to the list. When I first heard the call sign, "Cajun" I surmised that Maj. Del Bagno was of Cajun heritage — like me — or went to school in southwest Louisiana — Cajun Country — or perhaps was fond of Cajun

[6] All grades are from time in service with the Thunderbirds, not current.

cuisine. Curious, I reached out to "Cajun's" sister, Jeannette, who graciously provided me with some revealing insight. I was not only wrong, it was far from what I could have ever imagined!

> **"I don't know all the specifics as to how they actually receive their call sign…I'm sure that it is some super secret fighter pilot thing!"**
>
> Jeannette Del Bagno Crowder

Jeannette said, "I don't know all the specifics as to how they actually receive their call sign…I am sure that it is some super secret fighter pilot thing!" She said that according to the other pilots in the squadron, Stephen would mumble on the radio and was difficult to understand over the background noise. This earned him the call sign, "Cajun" which is an acronym for:

Can't **A**nnunciate **J**argon **U**nder **N**oise

Well, there you have it, you can't make this up!

USAF Thunderbirds Museum

United States Air Force Thunderbirds Flight Helmets

10
Protection, Incorporated

If you have never heard of Protection, Incorporated, a contemporary television program may come to mind, or perhaps a sinister organization from a Marvel or DC Comics page. A reasonable guess, but it would be incorrect.

Protection, Incorporated (PI) of Pomona, CA was the brain child of Dr. Charles Lombard, and at one time, the leading provider of custom flight helmets — most notably, the USAF Thunderbirds.[1] Protection, Inc. started out as a customer of GENTEX, customizing flight helmets for discerning clients in the USAF, U. S. Navy and NASA.

There is very little detail about PI except for those fading memories from retired military pilots, senior flight helmet collectors, scarce documents or what can be gleaned from the occasional helmet that shows up with PI markings.

Fortunately, we can get a little insight from the following statement of Company Qualifications[2] in an undated archival document:

"Protection, Inc., is a corporation organized and existing under the laws of the state of California. The company was originally organized in 1947 as a California corporation and has a continuous history since that time in the design, development and production of aircrew and personal protective equipment. In the period from January 1, 1957 to September 30, 1960, Protection, Inc. was operated as a division of Mine Safety

---

[1] Protection, Incorporated also made custom flight helmets for the U.S. Navy Blue Angels.

[2] From the personal archives of Mike Baetslé, Belgium

Appliance of Pittsburg, PA. As of October 1, 1960, its status was changed to a privately-owned California corporation and has continued as such to date."

In 1969, the custom-fit helmet business of Protection, Inc. was purchased by GENTEX. In 1983 GENTEX sold the business to Flight Suits, Ltd., now doing business as Gibson & Barnes.

Protection, Inc. had the Thunderbirds helmet contract from 1955 until 1983, when Flight Suits, Ltd. purchased the PI business. Currently doing business as Gibson & Barnes, the company hired only one Protection, Inc. employee and they also acquired the Thunderbirds flight helmet contract. The first Thunderbird flight helmets made by Gibson & Barnes were in 1984 for the 1985 show season.

1956, A group of industrialists and USAF experts with USAF test pilot George Smith (right).
l-r: Herman Roth, Pete Barker, Dr Charles Lombard, LtCol John P. Stapp and Smith discussing a new type of pilot's flying helmet. USAF pilot George Smith's experiences in being the first man to succeed in being ejected at supersonic speeds on 26th February 1955 and live supplemented and reinforced the experimental results.

Below is a copy of a GENTEX Corporation — Western Operations / Protection, Inc. flight helmet price list from 1982[3], which pre-dates the PI purchase by Flight Suits, Ltd. in 1983. Note the price difference between a basic HGU type helmet and the TOPTEX. As the saying goes, you get what you pay for!

Gentex®

PRICE LIST

CORPORATION—WESTERN OPERATIONS

2824 Metropolitan Place • Pomona, CA 91767 • (714) 596-6512

PROTECTION, INC.

Commercial

Effective July, 1982

CUSTOM FITTED HELMETS

Basic HGU Type Helmet — includes shell, visor housing with visor, liner edgeroll, earpads, one skull cap, and chinstrap. Helmet painted one color. Name on back of helmet.

Single Visor Helmet
(HGU-33/P) .... $ 445.00

Dual Visor Helmet
(HGU-26/P) .... $ 495.00

TOPTEX (or Lombard) Type Helmet — includes Kevlar® helmet shell with neutral visor, one skull cap, nylon fabric lens cover, and helmet bag. Helmet painted one color. Name on back of helmet. .... $ 575.00

Basic SPH Type Helmet — includes shell, visor housing with visor, liner, color coordinated retention system, Sonex® earcups (without earphones), and chinstrap. Helmet and visor housing painted one color. Name on back of helmet.

Single Visor Helmet
(SPH-4) .... $ 450.00

Dual Visor Helmet
(SPH-3B) .... $ 490.00

Special paint designs must be priced individually at the time the order is placed due to the individuality and complexity of the design.

Use of reflective tape and special paints must be priced at the time of the order.

3 From the personal archives of Mike Baetslé, Belgium

In another archival document from the personal collection of Mike Baetslé, we get a glimpse into a few of the major projects Protection, Inc. was working on in the mid to late 1950s.

12 JUL 1954	Wright-Patterson AFB Ohio	High Pressure Oxygen Helmet
9 APR 1957	Bureau of Aeronautics Washington, DC	Design and Improve Features of the APH-5 Helmet
26 DEC 1957	Holloman AFB New Mexico	Chimpanzee Protective Protective Garment[4]
24 DEC 1958	Wright-Patterson AFB Ohio	Protective Helmets (Redesign of HGU-2/P)
1 JUN 1960	Wright-Patterson AFB Ohio	Four Helmets - Wind-blast, Protective (Super TOPTEX, Model ST-7)

The previous list is impressive, to say the least, and underscores that fact that Protection, Inc. was on the cutting edge of protective flight gear technology. It's not surprising that PI was selected by the USAF Thunderbirds as their supplier of choice. That said, let's take a closer look at this historic partnership.

On the following page is a very rare letter[5] dated 27 January 1959 from Major Jacksel M. Broughton, to Protection, Inc. (misspelled "Protective") inquiring about ordering 20

4 Chimpanzee launched into space on January 31, 1961 as part of America's space program at the Holloman Aerospace Medical Center, located at Holloman Air Force Base in New Mexico,

5 From the personal archives of Mike Baetslé, Belgium

helmets, presumably, TOPTEX. This letter was written 2 years following Maj. Broughton's service as the Thunderbirds' second Leader from 1954 to 1957.

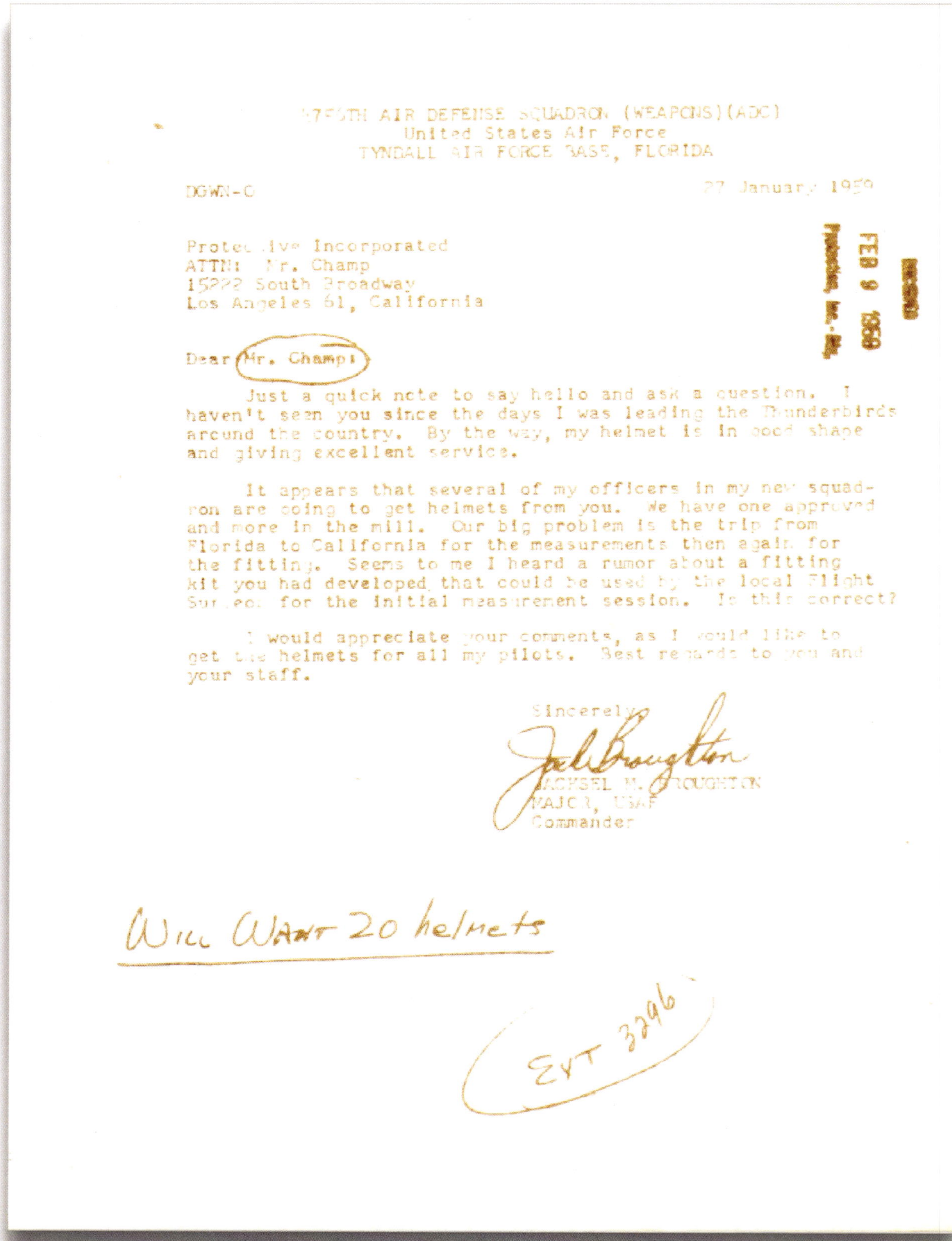

4756TH AIR DEFENSE SQUADRON (WEAPONS)(ADC)
United States Air Force
TYNDALL AIR FORCE BASE, FLORIDA

DGWN-C

27 January 1959

RECEIVED FEB 9 1959

Protective Incorporated
ATTN: Mr. Champ
15222 South Broadway
Los Angeles 61, California

Dear Mr. Champ:

Just a quick note to say hello and ask a question. I haven't seen you since the days I was leading the Thunderbirds around the country. By the way, my helmet is in good shape and giving excellent service.

It appears that several of my officers in my new squadron are going to get helmets from you. We have one approved and more in the mill. Our big problem is the trip from Florida to California for the measurements then again for the fitting. Seems to me I heard a rumor about a fitting kit you had developed that could be used by the local Flight Surgeon for the initial measurement session. Is this correct?

I would appreciate your comments, as I would like to get the helmets for all my pilots. Best regards to you and your staff.

Sincerely,

Jack Broughton

JACKSEL M. BROUGHTON
MAJOR, USAF
Commander

Will Want 20 helmets

EXT 3296

In an unrelated letter dated 11 April 1960 from Major Robert W. Casey, Sr.[6] to Protection, Inc., Maj. Casey is informing Mr. George W. Champe of his unsuccessful attempt to procure TOPTEX flight helmets for the 317th Fighter Interceptor Squadron. Maj. Casey writes, "...we gave it the old college try, but we were shot down."

"We gave it the old college try, but we were shot down."

Major Robert W. Casey, Sr.
317th Fighter Interceptor Squadron

It appears that PI sent Maj. Casey an example of a TOPTEX helmet for test, evaluation and presumably to assist him in his efforts to sway Air Force "brass" to approve his non-government procurement request. Maj. Casey said this of the sample helmet: "Got the helmet thru [sic] the mail a few days ago and as you say, it is a beaut. I still hope for some future success from demonstration and display of the little gem. It is a perfect fit and I now throw rocks at the P-4 type bucket."

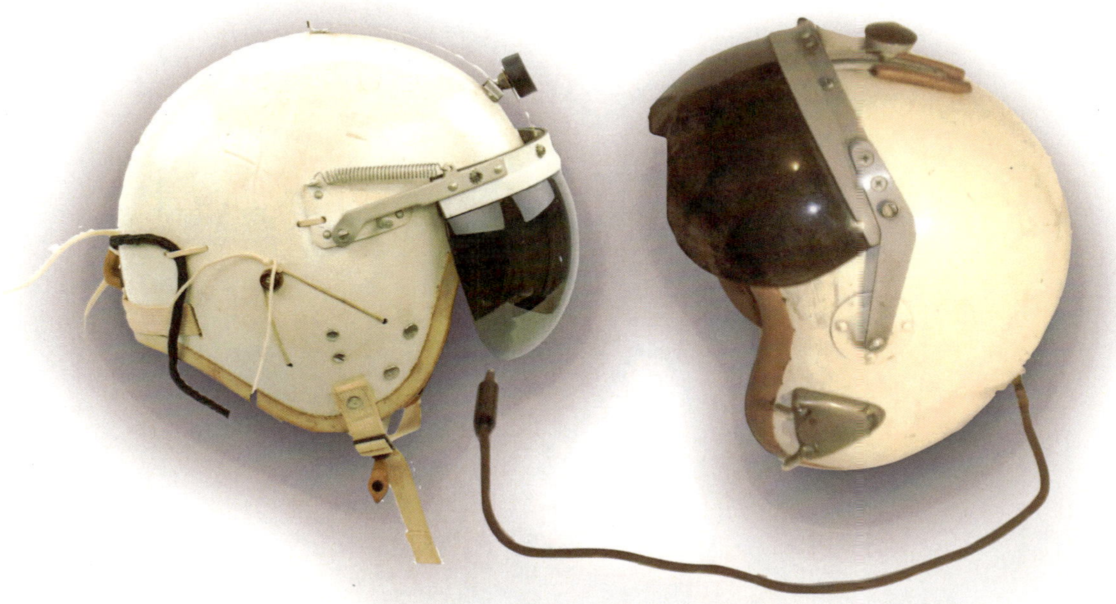

Standard USAF issue P-4 on the left; Protection, Incorporated TOPTEX on the right.

[6] From the personal archives of Mike Baetslé, Belgium

The image on the right is a PI tag from the interior of a USAF Thunderbirds F-4 "Phantom" flight helmet, which the team flew from the 1969 to 1973 show seasons. Although difficult to read, the tag bears the following information: NAME: Nellis AFB; DATE: (unreadable); MODEL: HGU-2A/P; LINER NO: NO LINER.

Several fellow flight gear collectors from across the globe were kind enough to share several PI helmet decals from their personal collections. These tags are located on the inside of the helmet shell, behind the leather liner.

The next tag (top to bottom) is from Salvo Zannelli from Italy.

The second two tags are from the personal collection of Laurent Michelot, a fellow flight gear collector from France. It's interesting to note that these tags are dated 1979 and 1981, and are still being used many years after GENTEX acquired PI in 1969.

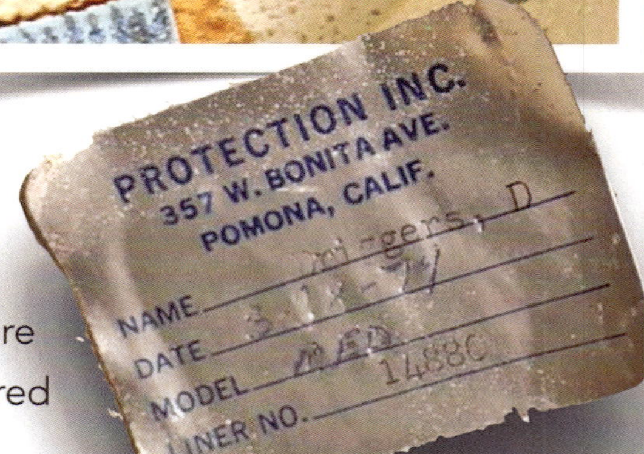

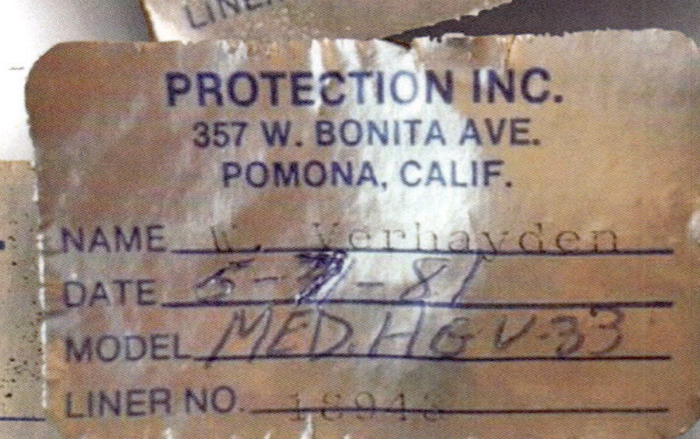

Marc Koch, from Luxembourg also shared a PI tag from his private helmet collection. (bottom)

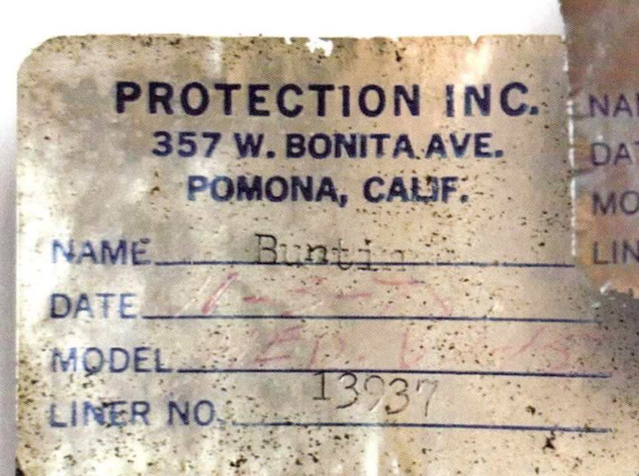

Besides the customer tags, PI "branded" the aluminum "diamond" plates on the oxygen mask bayonet receivers as a means of identifying and advertising their products. PI was also well known for their use of chrome plated receivers, as evidenced by the photos on the right.

In some cases, Protection, Inc. would also hand-mark the inside of the visor housing, presumably a means to identify a customer's helmet during the painting and assembly process.

The photo on the bottom is an example of a strangely familiar yellow visor housing from P.I. that was for a rival, albeit friendly competitor team!

PROTECTION, INC. DIVISION

TOPTEX HELMET

Manufactured by:

PROTECTION, INC. DIVISION
15222 So. Broadway
Los Angeles 61, California

INSTALLATION AND ADJUSTMENT
INSTRUCTIONS

TOPTEX OXYGEN MASK RETAINER KIT, PART NO. H3740

GENERAL

The Toptex oxygen mask retainer kit is designed to afford improved retention of the helmet-mask combination when exposed to high wind-blast forces. It consists of the following:

(1) A molded fiber-glass retainer shield (available in three sizes to match the three sizes of the A-13A oxygen mask), provided with an attached strap system terminating on each side in an adjustable bayonet; and (2) two latch assemblies, one for each side of the helmet, into which the bayonets are inserted. Four sets of notches on the bayonet enable adjustment of mask pressure against the face, and spring-loaded release levers allow selective reduction of this pressure, or complete release of the bayonet when removal of the mask is desired.

The base of the latch housing is curved to adapt to the contours of principal types of helmets now in use, and a backing plate is provided to afford good load distribution on the inside of the helmet.

The following instructions cover both installation of the kit on the Toptex helmet, and adjustment to the individual wearer. When latches have been factory-installed, refer only to the sections on mask modification and adjustment to the wearer. The assistance of someone trained in oxygen equipment fitting and maintenance is urgently recommended.

TOOLS AND SUPPLIES NEEDED FOR MODIFICATION AND INSTALLATION

Pocket knife; 4 to 6 inch screwdriver for the type of screw provided; end cutters; needle and No. 10 thread; sewing machine; four battery clamps or their equivalent; No. 25 drill; 1/4" electric hand drill; and a 6-penny nail, for marking location of holes to be drilled in the shell. (The last three items are not needed if latches have been installed at the factory.)

MASK MODIFICATION

1. Make sure that the mask and shield are matched for size (i.e., small, medium or large; shield size is marked on its interior surface). Remove all straps and cut off the protruding rubber strap attachment loops from the mask by cutting flush at the mask surface with a sharp knife or a pair of end cutters. Discard the plastic piece surrounding the two top strap loops. Cut off the rubber cross strap between the sealing flanges, since this will no longer be required to prevent mask spreading due to either pull-down on the

2500 3-57 T-3143

PROTECTION, INC. DIVISION

face, or higher breathing pressures.

HELMET MODIFICATION (PROTECTION, INC. DIVISION TOPTEX HELMETS)

With the earphone pads removed from the helmet, remove the rivets holding each oxygen mask tab to the shell (drill out with No. 25 drill or cut with heavy-duty end or side cutters). Pull the tab away from the shell (the tab was originally cemented as well as riveted). Remove any other accessories, such as boom mike, if located in the latch mounting area shown in Figure 1. (Special mounting will be required for boom mike if this item is needed.)

INSTALLATION

Before installing the latches, try inserting and releasing the bayonet to become familiar with the operation. Note that the bayonet has four sets of notches and therefore four stepwise degress of insertion. Note that both release levers have to be depressed to release the bayonet, and also, during insertion, that both levers should "click" into their respective notches for maximum holding power.

Remove the backing plate from the latch assembly as shipped, noting that short attaching screws are used on the "back" side of the body, where the screws are closer together, and longer screws on the front, where the release levers extend from the central channel.

In mounting the latch on the Toptex Helmet, one of the three holes formerly used to secure the oxygen mask tab (the top front hole) can serve as the top front mounting hole for the latch. Place one long screw with lock washer through the top front hole of the backing plate, and from the inside, push the screw through the specified mounting hole in the helmet. Start the screw into the top front mounting hole of the latch housing, and tighten with the lower front edge of the latch body 3/4 inch back from the line along which the leather edge roll is attached to the shell. (See detail "B", Figure 1.)

Repeat the operation with the latch on the other side, and check both sides to see that the latches are "aimed" symmetrically. This latch position has proved satisfactory in the majority of cases, but prior to drilling the rest of the mounting holes, it is preferable to proceed with the next step. Before the helmet is put on, make sure that the single mounting screw holding each latch at this stage is firmly tightened.

ADJUSTMENT TO WEARER

Place the stripped oxygen mask into the retainer shield assembly, and then insert the left bayonet into the left latch (left being in relationship to the wearer). Insert to the third notch position,

- 2 -

counting from the first "click" as the bayonet is inserted. Put the helmet on the head and insert the right bayonet into its latch, also in the third notch position. Adjust the straps for normal wearing tension of the mask against the face.

To hold the straps during preliminary adjustment, medium sized battery clips are most convenient to hold the looped ends, but masking tape can be used as a substitute. In adjusting the straps, aim for a good balance between top and bottom strap tension, and also make sure that the mask is centered over the nose with the helmet in normal position on the head. It is advisable at this point also to check with various oxygen pressures in the mask.

Due to variations in facial contour, it may be desirable in some cases either to raise or to lower the average angle of pull toward the face. This can be done by changing the latch housing position. The tension angle can be <u>raised</u> by moving the lower edge of each latch housing <u>forward</u>, and <u>lowered</u> by moving the lower edge of each latch housing <u>backward</u>, with the top still held by the one mounting screw. Strap readjustment during this process will also be necessary.

When a satisfactory adjustment has been achieved, secure the latches by the rest of their mounting screws. Hole locations for the mounting screws can be indicated on the shell by marking through the mounting holes in the latch housing (a 6-penny nail can be inserted and tapped lightly). With the housing swung out of the way, the holes can then be drilled using a No. 25 drill. Insert the mounting screws with their lock washers from the inside and tighten gradually and successively until the latch housing is snugged down against the helmet shell. Avoid excessive torque on the screws.

The straps should then be machine sewn at the desired adjustment using No. 10 thread or approximate equal. Refer to sewing detail, Figure 1, for the recommended pattern for sewing both top and bottom straps. Any excessive length of strap beyond the sewn area can be cut off. Heat sealing of cut end of strap is advised.

Kits can be supplied with buckles on the bottom straps to facilitate readjustment in the event of re-assignment of helmet equipment. Since most equipment is personalized and permanently assigned, this buckle can be omitted and machine sewing of all straps is recommended.

When the mask has been adjusted for normal wearing tension with the bayonets in the third notch position on each side, <u>two</u> additional steps of increased tension are actually available, since each bayonet can be tightened independently another notch.

If the helmet has a visor, check the fit of the lower border of the lens to the mask retention assembly, and trim the lens carefully if interference prevents proper locking of the visor in the down position. Visor trimming should be done by someone skilled in handling acrylic plastic materials.

OXYGEN MASK HOSE NON-STRETCH

Recent experience has demonstrated that under today's high wind-blast forces during seat ejection, the oxygen mask inlet hose and its connection to the mask often fail. Use of a nylon-covered, non-stretch hose with a metal reinforcing ferrule and stiffener ring is recommended.

INSTALLATION ON OTHER HELMETS

Although the Toptex oxygen mask retainer was designed for and will give best results with the Toptex helmet, it can be used also with other helmets having adequate shell strength. Proper latch location on such helmets can be determined by application of the general principles set forth in the foregoing instructions.

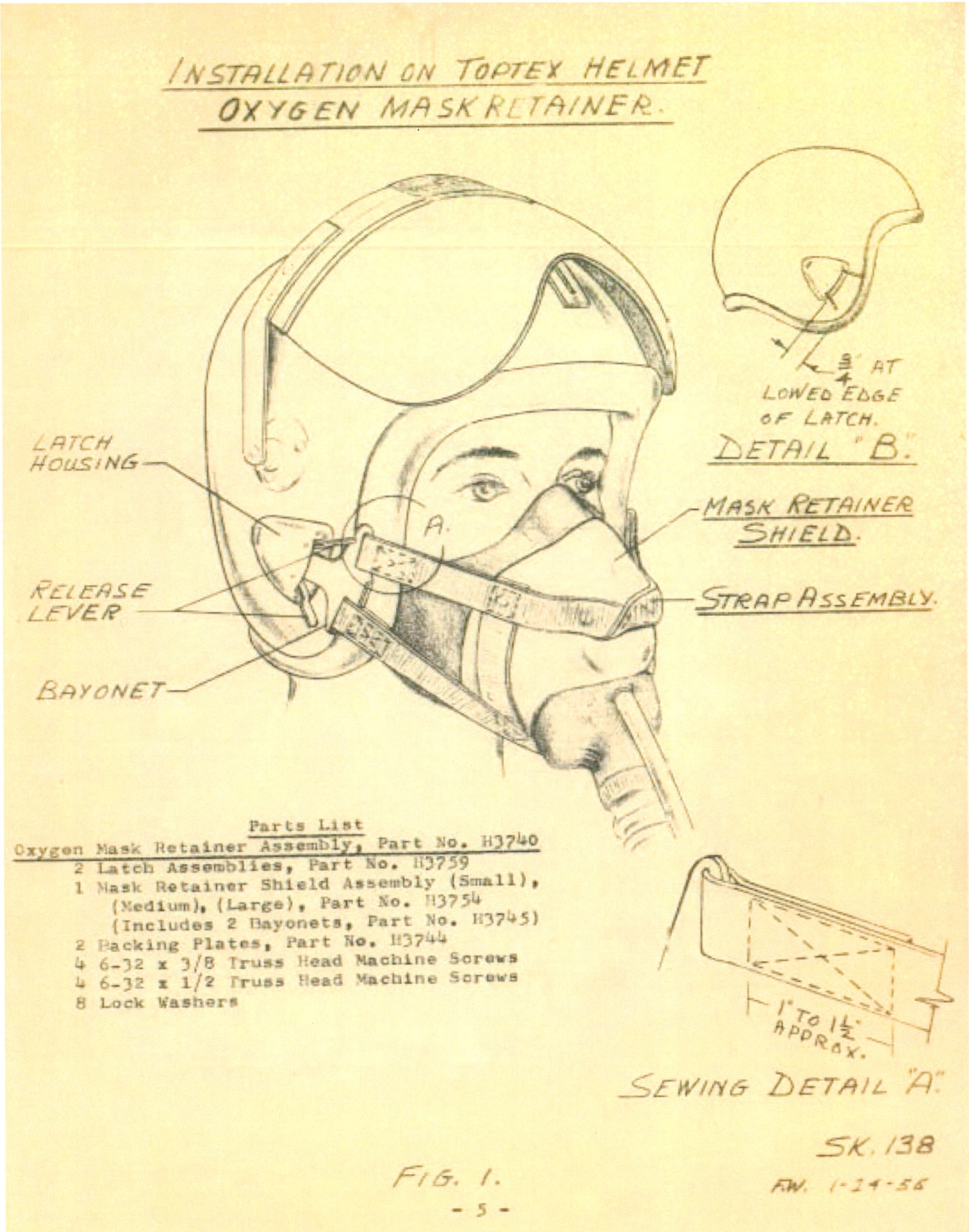

Parts List

Oxygen Mask Retainer Assembly, Part No. H3740

2 Latch Assemblies, Part No. H3759
1 Mask Retainer Shield Assembly (Small), (Medium), (Large), Part No. H3754 (Includes 2 Bayonets, Part No. H3745)
2 Backing Plates, Part No. H3744
4 6-32 x 3/8 Truss Head Machine Screws
4 6-32 x 1/2 Truss Head Machine Screws
8 Lock Washers

- 5 -

PROTECTION, INC. DIVISION

VISOR INSTALLATION INSTRUCTIONS

The following instructions apply to installation of this visor on Toptex Helmets. CAUTION: Leave protective film on visor lens during installation and until lens is trimmed to integrate with oxygen mask.

Figure 1 is a view showing the proper relative positions of latch plate and pivot assembly on helmet.

Referring to Figure 2, locate mounting holes by using the following procedure:

1. On one side place a straight edge against the leather covered edge roll on the outside where it meets the shell surface in the cheek area.

2. With a pencil draw a line approximately 2 inches in length upward from start of edge roll radius.

3. Repeat on opposite side.

4. Using the upper side of the edge roll in front as a beginning, draw a line parallel with it along the sides several inches beyond the vertical lines described in Nos. 2 and 3.

5. Along the line at the upper side of the edge roll measure the distance between the vertical lines and bisect it to get the front center indication.

6. Measure the distance over the top between the two horizontal lines and bisect it to get the top center indication.

7. Using the front and back center indications, draw a line between them. (This is the helmet center line.)

8. Indicate a point 1-1/2 inches and another 6-1/2 inches up and on line 7.

9. Take a pair of dividers or a compass and set at 6-1/4 inches. With one leg of the compass placed on the 1-1/2 inch point in 8, scribe an arc on each side of the helmet in the area where the pivot assembly is to be located. Do the same at the 6-1/2 inch indication. This will give the center location of the pivot assembly.

10. Center the locking holes of the latch plate over the 1-1/2 inch and 6-1/2 inch indications. Locking hole nearest to the end is towards the front. Holding the latch plate in place, indicate

- 6 -

2500 3-51 T-3148

the 4-40 screw holes on shell with a sharp pencil or small center-punch.

11. Drill with No. 30 drill in both sides, center punch and drill hole "C" (pivot axis) with a No. 30 drill for pilot hole and redrill with a No. 10 drill.

12. Position the center hole of the pivot assembly directly over hole "C". Turn the pivot assembly until short side is toward bottom and rear of helmet as shown in Figure 1. NOTE: This position is important. The two lower holes should be parallel to the line running along the top edge of the edge roll.

13. Drill four rivet holes with No. 30 drill with the pivot assembly serving as drill jig. Be sure pivot assembly remains positioned throughout the drilling operation.

14. Rivet pivot assemblies to shell with rivets enclosed using washers on the inside of the shell.

15. Install the latch plate loosely. Install visor lens assembly on pivots using the 10-32 clevis bolts provided. Screw 10-32 screws two to three turns only.

16. Test entire installation for alignment before tightening screws. If misaligned, file holes in shell to one side or the other until aligned.

ADDITIONAL INFORMATION

1. A small plate has been added to the upper rear underside of the latch plate. This prevents the latch pin from engaging securely.

2. Once the latch plate has been installed, use the front locking hole as a guide, and drill through the fiberglass shell to enable the locking pin's full engagement. Use a No. 10 drill.

CAUTION: Do not attempt to pull visor down until outer edges have been trimmed so as not to damage finish on edge roll.

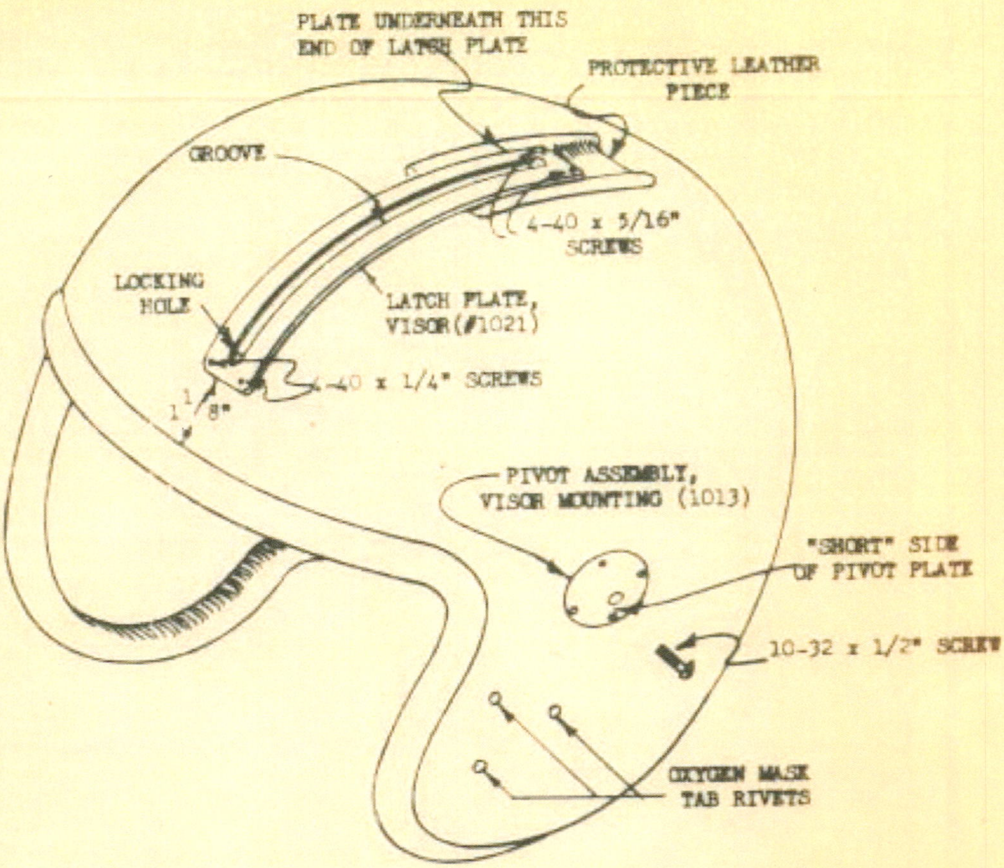
PLATE UNDERNEATH THIS
END OF LATCH PLATE
PROTECTIVE LEATHER
PIECE
GROOVE
4-40 x 5/16"
SCREWS
LOCKING
HOLE
LATCH PLATE,
VISOR(#1021)
4-40 x 1/4" SCREWS
1 1/8"
PIVOT ASSEMBLY,
VISOR MOUNTING (1013)
"SHORT" SIDE
OF PIVOT PLATE
10-32 x 1/2" SCREW
OXYGEN MASK
TAB RIVETS

Fig. 1

- 8 -

PROTECTION INC.

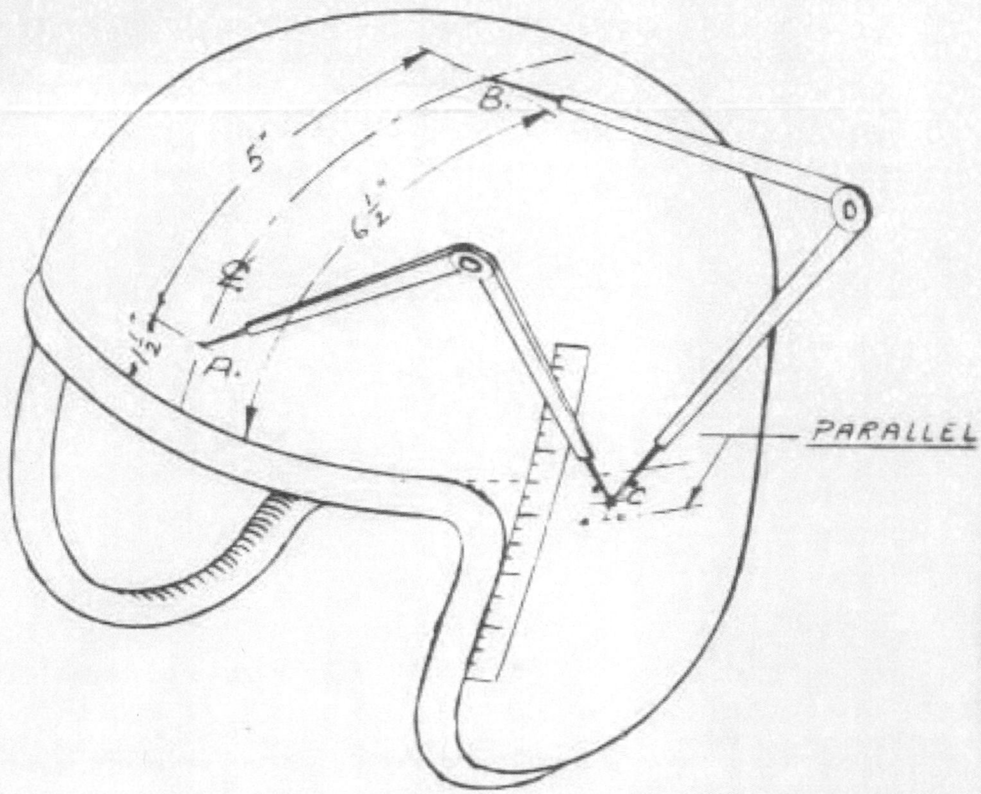

FIG. 2

- 9 -

TRIMMING:

CAUTION: Leave protective coating on lens until trimming has been completed.

1. Place helmet with visor on head, install oxygen mask, adjust to normal wearing tension, and position.

2. Lower the visor lens until it contacts the mask. Mark the line of contact on mask at several points - at least at top center, and two points on each side. Then raise visor.

3. Using heavy cardboard, cut to contour of mask along the visor contact line.

4. Lower lens again until it is in contact with the mask. Then measure the distance between the latch pin and the lower detent hole of the latch plate. Deduct 1/8 inch from this distance, giving a dimension which will be referred to below as "A". See sketch on page 12.

5. Measure upward by dimension "A" from the center of the mask where the lens will contact, and mark a point on the lens.

6. Remove helmet and visor from head, and take visor assembly from helmet.

7. Place cardboard template against lens with the center top of the mask cutout area at the point which was located in Step 5, and evenly spaced from the yoke, on each side. Use wax crayon (or china marking pencil) and mark along lower edge of template.

CAUTION: Leave the lens as deep as possible on each side of mask, so as to minimize any gap between the lens and the face.

8. Cut along outer side of line thus marked, using band saw, jig saw, or coping saw.

9. Before going further, replace visor on helmet (pivot screws do not have to be turned down tight), and determine whether the trim is approximately correct. Mask should be worn with normal tension. Lower the lens and see whether edge contacts mask evenly. If not, mark points where additional trim will be necessary. Also check to see whether latch pin will drop into locking hole with slight pressure of lens edge on mask.

10. When proper fit has been obtained, dress off edge of lens with file or sandpaper. The dressed edge should not be square with the surface of the lens, but should have an angle which will match the surface of the mask at the contact line. Also, round all sharp edges and make finished surface as smooth as possible. Do not buff

since this can easily overheat edge and cause trouble.

11. Finally, remove protective coating from lens.

PROTECTION, INC. DIVISION

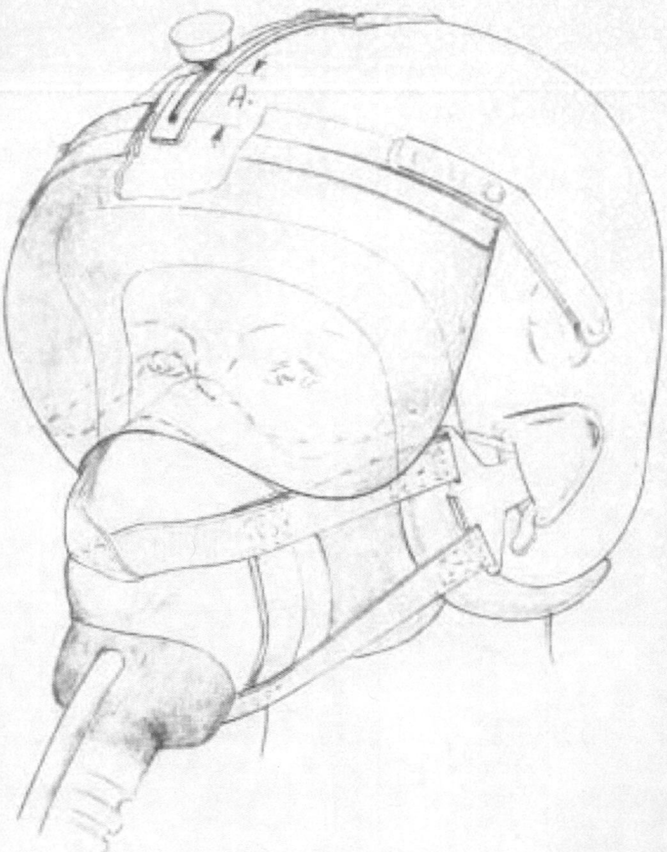

Dotted Line shows approximate trim.

- 12 -

PROTECTION, INC. DIVISION

INSTALLATION INSTRUCTIONS

FOR

NYLON COVERED NONSTRETCH HOSE

Remove hose and fiber ferrule from mask.

Remove bailout adaptor from hose

Put slotted stiffener ring in groove on mask. Push metal ferrule into mask replacing fiber one. Slip on nonstretch hose making sure inner ridge seats firmly into stiffener ring. Replace Tinnerman clamp and rubber microphone wire holder.

On opposite end of hose, replace bailout adaptor and Tinnerman clamp.

In the preceding document from Protection, Inc. Division, we are afforded a rare and comprehensive look into the installation and adjustment instructions for the TOPTEX flight helmet. Although the document is undated, one of the drawings is dated 1956, providing an important time reference. This document is courtesy of and from the personal archives of Mike Baetslé, Belgium.

The Protection, Inc. Division document on the following page (also courtesy of and from the personal archives of Mike Baetslé, Belgium) details the care and maintenance of flight helmet molds.

"...it would be wise to make the molds as light as possible...to realize that girls and not men are going to have to handle them."

Dr. Charles F. Lombard
TOPTEX Memo
02 May 1958

In a separate document dated 1958, Dr. Charles Lombard writes, "Likewise, if at all possible, it would be wise to make the molds as light as possible by not making the sides too thick (over 3/8 inch). This last thought is to make the molds strong enough but to realize that girls and not men are going to have to handle them."

Two additional Protection, Inc. Division documents, both dated 1958, are found on subsequent pages that reference Standard TOPTEX Helmet Shell Tooling, and Standard TOPTEX Helmet Shell Molds.

These documents were also from the personal collection and provided through the generosity of Mike Baetslé, Belgium.

PROTECTION, INC. DIVISION

WRITTEN STANDARD PRACTICE

Date June 19, 1958

By E. B. Kwiatkowski

No. 6

CARE AND MAINTENANCE OF MOLDS

The following procedure is to be followed to maintain all molds in good condition:

1. Molds shall be washed once per week with solvent to remove all mold release.
2. Wipe molds dry.
3. Clean off any excess resin.
4. Clean out all bolt holes.
5. Apply a heavy coat of Trewax.
6. Polish.
7. Apply a second coat of Trewax.
8. Polish.
9. Apply one coat of mold release paste 550.
10. Polish.
11. Apply one brush coat of mold release #3, making sure all parts of mold are coated.
12. Reassemble mold. Mold is now ready for use. After removing shells, recoat mold with one coat of mold release #3 each time mold is used between cleanings.
13. Report to Supervisor if molds show any signs of damage or wear.

W.S.P. No. 6

2500 3-57 T-3182

CONFERENCE REPORT June 17, 1958

PRESENT: L. P. Corbett
G. W. Champe
J. C. Miller
Ed Kwiatkowski
H. P. Roth

Subject: Standard Toptex Helmet Shell Tooling

This conference was concerned primarily with the problem of mounting the visor on the helmet shell in correct geometrical relation thereto. It was decided that the mold for each type of shell would be modified as follows:

1. The proper location for the pivot of the visor on each side will be determined and transferred to the interior of the mold.

2. The mold will be drilled at that point, and a metal pin embedded in such a way that a corresponding dimple will be molded into the side of each shell as it is laid up and cured.

3. In addition, a pin will be embedded to indicate on each side of the shell the location of one of the four mounting holes of the pivot plate.

4. If there is more than one mold for a given shell type, all such molds will be examined and the best one selected for use in regular production. Use of the others will be avoided.

This work will be done by Kwiatkowski and Roth on a Saturday basis.

Herman P. Roth

HPR:br

COPY

MEMORANDUM June 15, 1959

TO: H. P. Roth

FROM: E. B. Kwiatkowski

SUBJECT: Standard Toptex Shell Molds

3AM - old mold - no new molds made. This shell is seldom used because of the very wide neck opening.
Length - 10" Width - 8-7/8" Neck - 7"

3SM - old mold dimensions:
Length - 10-1/16" Width - 8-7/8" Neck - 6"
Depth - 9-1/8"

3SM - new mold dimensions:
Length - 9-13/16" Width - 9-1/8" Neck - 5-7/16"
Depth - 9-7/16"

3BM - old mold dimensions: 2 each
#1 - Length - 10-3/16" Width - 8-5/8" Neck - 5-3/8"
#2 - Length - 10-3/16" Width - 8-7/8" Neck - 5-9/16"

3BM - new mold dimensions:
#1 - Length - 10-1/8" Width - 8-7/8" Neck - 5-3/8"
#2 - Length - 10-1/4" Width - 8-7/8" Neck - 5"

3BL - mold is in development department for redesign of face opening.

Frequency of use is as follows:

BM	100 approximately
SM	30 "
AM	7 "
BL	2 "

More BM shells will be used than the above figures show due to the change in head size range for the SM shell.

EBK:hep

E. B. KWIATKOWSKI

PROTECTION, Inc. Division

The Protection, Inc. documents in this chapter are unquestionably, treasured historical artifacts that continue to tell a story, despite the passing of six decades. The ink has faded but the recorded history on those yellowed pages remains sharp and clear. We can't query the letters, memoranda and reports, yet they answer many questions. Short of visiting a business that closed its doors over 40 years ago, or speaking with former employees, these company files are the best we can hope for to capture and preserve an important part of the USAF Thunderbirds flight helmet lineage.

Officially, the company name was Protection, Inc., a Division of Mine Safety Appliance (MSA). Unofficially, they were known as P.I. Whatever you prefer to call them, they were the first, and without question, one of the premier suppliers of custom USAF Thunderbirds flight helmets from the late 1955 until 1983. That period includes the TOPTEX helmets for the F-100C "Super Sabre", F-105B "Thunderchief" and F-100D "Super Sabre" eras, as well as the HGU-26/P helmets for the F-4 "Phantom" and T-38 "Talon" eras. In a word, impressive.

In the flight gear collector's world — of which I am a card carrying member — flight helmets produced by Protection, Inc. are highly sought after for their quality and beauty. Add to this their scarcity, and these helmets routinely command prices two to three times what a helmet produced by GENTEX or Sierra Engineering might bring through an online auction. In the world of economics, low supply and high demand equates to big bucks!

Without question, Protection, Inc. has earned a place of honor in the history history books. P.I. not only produced custom flight helmets for the USAF Thunderbirds and the U. S. Navy Blue Angels, but they were the preferred choice by military test pilots and NASA.

As long as there are flight gear collectors, Protection, Inc.'s legacy will live forever!

Lucien Van Oosten
Protection, Inc. / GENTEX

Mark Jones, Flight Helmet Manager for GENTEX, shared with me that a former Protection, Inc. employee was still employed by GENTEX in Southern California. Mark introduced me to Lucien Van Oosten, who is currently working as a Senior Engineering Specialist. I was privileged that Mr. Oosten accepted my request for a telephone interview.

Mr. Oosten began working for Protection, Inc. in 1981, and transferred to GENTEX when they purchased the custom fit helmet business from P.I. in 1983. While employed with P.I., Mr. Oosten worked on the flight helmet design engineering team. According to Mr. Oosten, P.I. — the brainchild of Dr. Charles Lombard and Bob Zimmerman — pioneered the custom fit flight helmet business for USAF test pilots, Lockheed, Douglas Aircraft and many more. Between 1981 and 1986, Mr. Oosten was actively involved with custom flight helmets, paint and graphics.

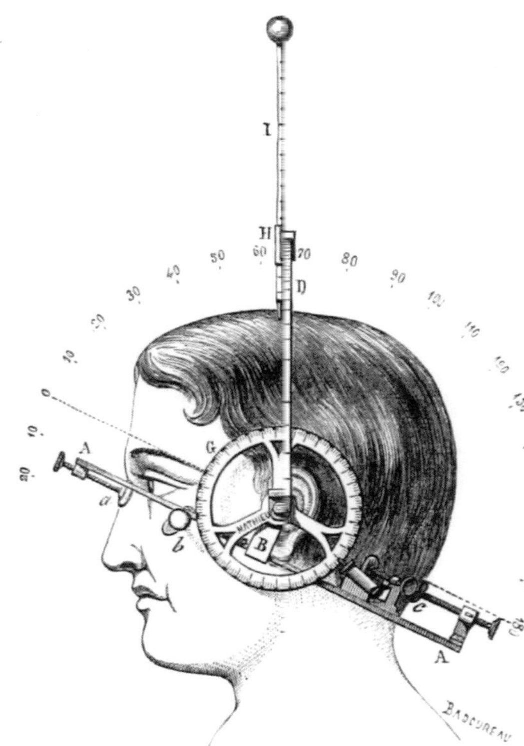

Fig. 45.

The conversation quickly turned to the TOPTEX helmet, discussing the success and ultimate demise of this flight helmet that was far ahead of its time. Mr Oosten says that P.I. focused a good portion of their research into anthropometry, the study and understanding of human physical variation, specifically, the systematic measurement of the physical properties of the human body. In the case of flight helmets, the primary focus was the size and shape of the head and face. P.I.'s research resulted in flight helmets that afforded the wearer better fit, comfort, stability, less fatigue and improved sound attenuation.

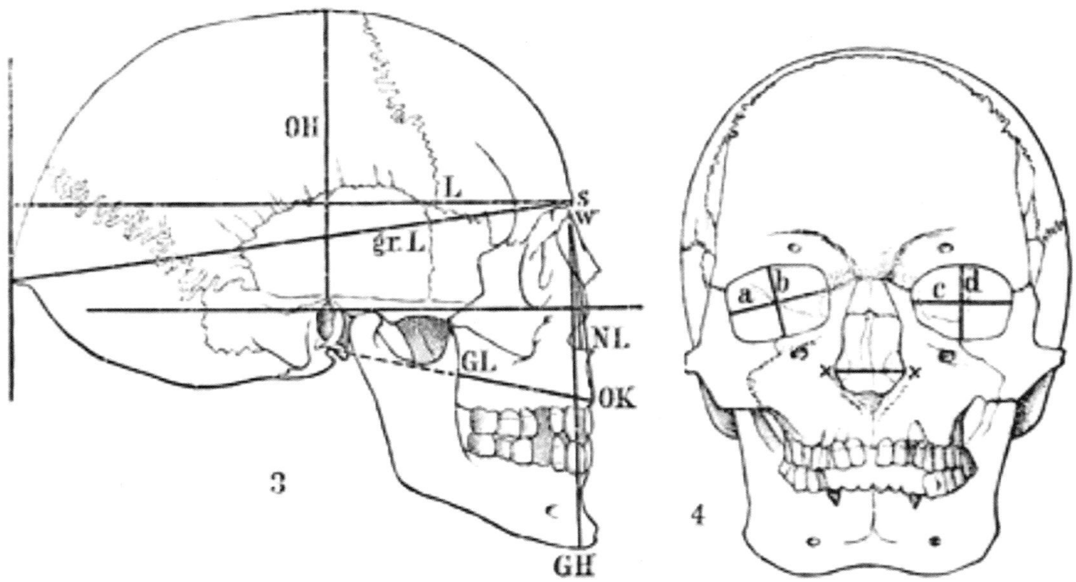

Unfortunately, these helmets required on-site, custom fittings that began with a mold placed on the pilot's head. The mold was filled with bee's wax softened to 130 degrees Fahrenheit which produced an exact copy of the pilots head. Custom fitted liners were then made from the molds. The time investment for each pilot being measured, and the economics of this expensive option to USAF issued flight helmets, led to the TOPTEX helmet fading quietly into the history books. A sad commentary on a flight helmet that was far superior to anything else the U. S. Government had to offer at the time.

In a 1959 dated letter from former Thunderbird Commander / Leader Jack Broughton, to Protection, Inc., he expressed his [TOPTEX] helmet was in good shape and continuing to provide excellent service. In his assignment following the Thunderbirds, Major Broughton was interested in obtaining 20 TOPTEX helmets for his squadron; however, he was concerned about each pilot having to make two trips from Florida to California, one for measurements and the second for fitting. Maj. Broughton inquired about a fitting kit that could be used in the field, but unfortunately, we are unaware of the reply. We do know that years later, a similar type of fitting mold was developed for USAF issued flight helmets that could be used by Life Support Specialists at the squadron level.

According to Mr. Oosten, the USAF issued flight helmets at the time were very "spherical" in shape and have changed very little today. By contrast, the TOPTEX helmet was slightly narrower from ear to ear, and slightly "elongated" from front to rear. These differences were the result of their study and application of anthropometry. Science aside, the pilots lucky enough to be fitted and fly with a TOPTEX helmet were immediately impressed and often became repeat customers.

There are a few surviving documents that reflect the sentiment of some satisfied customers. In 1960, Major Robert W. Casey, Jr., 317th Fighter Interceptor Squadron wrote:

"It [TOPTEX] is a perfect fit and I now throw rocks at the P-4 type bucket."

In a separate 1960 letter, 1st Lieutenant Robert, P. Krueger, 523rd Tactical Fighter Squadron wrote:

"...back in February this year, six officers...visited your factory for the purpose of fitting, adjusting and picking up their new 'Toptex' flight helmets. We wish to report we have all received satisfactory results from the wearing of your helmets. None of us experienced any uncomfortable feeling due to the wearing of these helmets."

These are impressive testimonials. Unfortunately, we will never know what would have become of Protection, Inc and their TOPTEX helmet if funding or logistics would have been different. What we do know, is that despite the passing of 60 years, the TOPTEX is still considered one of the finest flight helmets ever made. Period.

11
GENTEX

All military and civilian flight helmets, whether fixed wing or rotor wing, prop or jet, begin their useful life in one of a very limited number of plants specializing in high tech protective headgear. This includes every USAF Thunderbirds flight helmet. Before the helmets receive their flashy paint and distinctive Thunderbirds graphics, each one is hand created from fiberglass or Kevlar and resin and molded under heat and high pressure inside the world's foremost flight helmet manufacturer: GENTEX Corporation.

The General Textile Mills - now doing business as GENTEX - is a privately held company that focuses on the manufacture of Personal Protective Equipment products for the U.S. and international military, special forces, commercial, law enforcement, emergency medical services and First Responder, as well as aluminized fabrics, Lifetex Fabrics, Clearweld, Filtron, and Precision Polymer Processors.

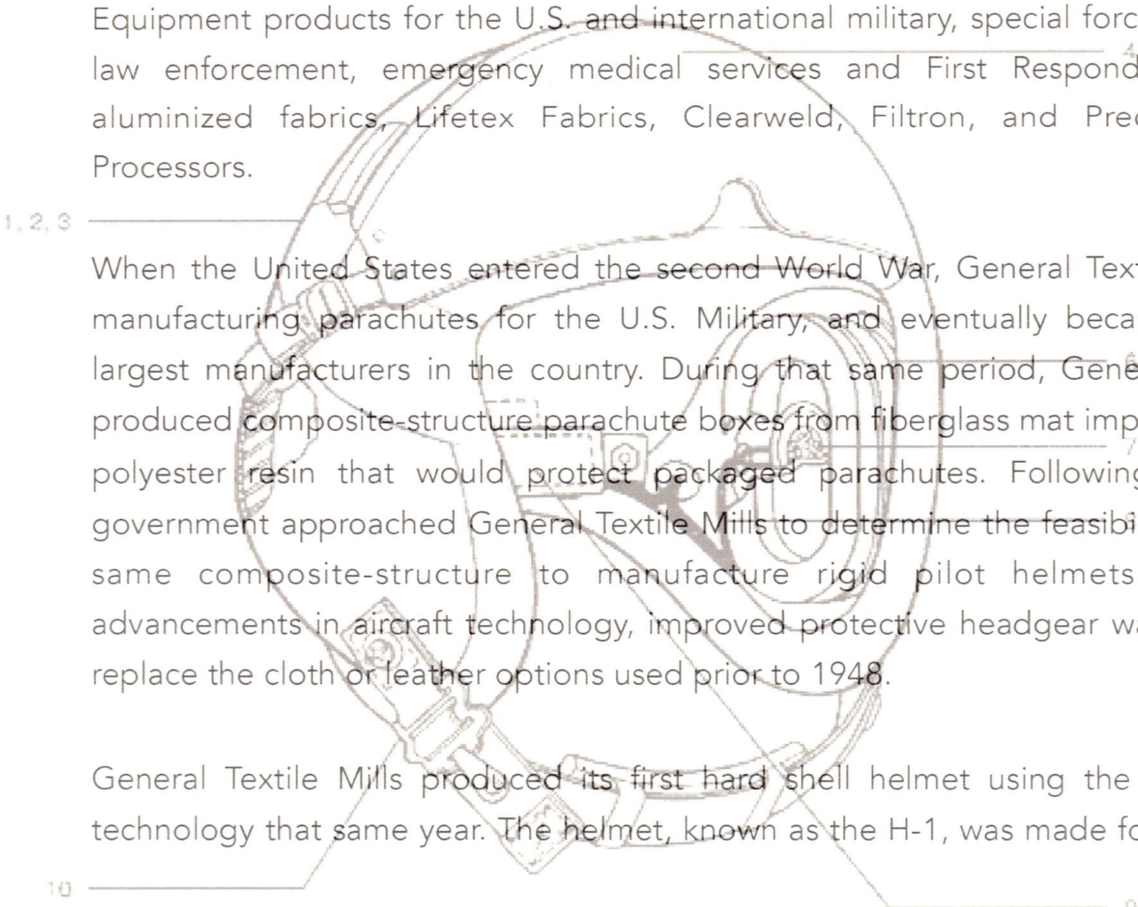

When the United States entered the second World War, General Textile Mills began manufacturing parachutes for the U.S. Military, and eventually became one of the largest manufacturers in the country. During that same period, General Textile Mills produced composite-structure parachute boxes from fiberglass mat impregnated with a polyester resin that would protect packaged parachutes. Following the war, the government approached General Textile Mills to determine the feasibility of using the same composite-structure to manufacture rigid pilot helmets. Due to the advancements in aircraft technology, improved protective headgear was warranted to replace the cloth or leather options used prior to 1948.

General Textile Mills produced its first hard shell helmet using the parachute box technology that same year. The helmet, known as the H-1, was made for the U.S. Navy and the P-1 soon followed for the United States Air Force. These technologies were quickly incorporated into helicopter pilot helmets for both military and commercial applications, as well as tank crew helmets.

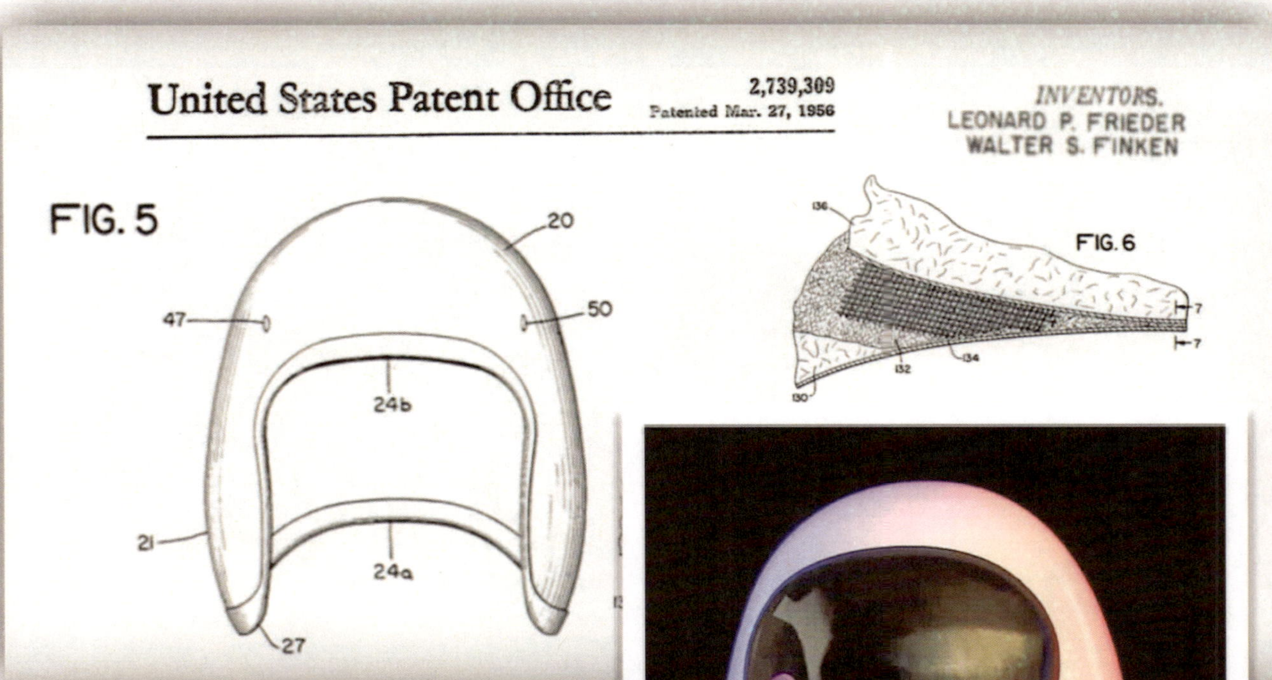

In 1958, General Textile Mills shortened its name to GENTEX, and the company continued working to improve and adapt the helmet technology.

U.S. Navy H-1 flight helmet manufactured by GENTEX.

Significant advancements in the aircraft industry resulted in the need for oxygen-breathing equipment for life support. In 1969, GENTEX bought Protection, Inc., a helmet manufacturer in Pomona, California. Ten years later, Protection Inc. began to acquire equipment in order to manufacture respiratory products. By 1981, the new branch officially transitioned into GENTEX Western Operations focusing on the manufacture of respiratory products. The California operation moved to Rancho Cucamonga in 1992.

GENTEX Corporation works closely with international distributors such as Transaero, Melville, New York, to provide products to international armed forces, and U.S. commercial distributors, including Flight Suits, Ltd., doing business as Gibson & Barnes, El Cajon, California. Plus, the company is called upon for specialized projects that require customization of personal protective products for NASA programs such as the early Space Shuttle missions and commercial projects including Burt Rutan's Scaled Composites and SpaceShipOne as part of the race for the Ansari X Prize.[1]

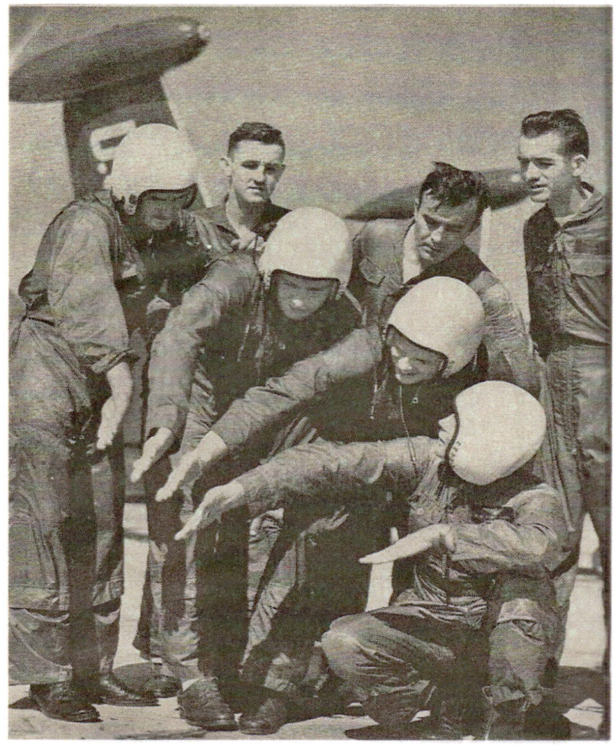

U. S. Navy Blue Angels wearing their GENTEX H-1 Flight Helmets. (*LIFE Magazine,* 5 DEC 1949)

In 1983, GENTEX sold the custom-fit helmet business, doing business as Protection, Inc. to Flight Suits, Ltd., doing business as Gibson & Barnes.

In the late summer of 2018, I wrote to GENTEX, seeking their support and assistance with my research for this book. I was promptly contacted by Mark Jones, Aircrew Helmets Product Manager, who became my primary point of contact, and over the next few months, provided me with in depth historical information, as well as arranged for a plant tour in Simpson, Pennsylvania.

[1] https://en.wikipedia.org/wiki/Gentex_Corporation

My tour was scheduled for the first week of December 2018. I was advised the formal portion of the tour would take approximately one hour leaving plenty of time to pick the brains of the "seasoned employees" on the production line.

The GENTEX Plant is located in an unassuming building and nestled among an established neighborhood on Main Street in Simpson, PA. Visitor parking is across the street next door to a craftsman style home typical of the area of less than 1,300 residents.

The tour began with a brief oral history of The General Textile Mills - GENTEX - just inside of the sprawling 100+ year old, multi-story plant. The wooden floors were original to the building, well worn, and scarred from several generations of families who have worked here since before World War One. I learned from talking with several employees that 40 to 45 years of service is not uncommon. A true testament to a successful, well run, family business that recognizes the importance of a loyal workforce. There is a sign proudly hanging in the plant that reads:

"GENTEX: Lives Depend on Us."

As we began what turned out to be a 2 1/2 hour tour, I was most surprised by the very limited use of automation and the lack of a "production line" that likely comes to mind when manufacturing anything in mass quantities. Instead, there were dozens of small, informal work stations scattered throughout the plant where hundreds of dedicated employees hand-build every single flight helmet.

The flight helmet process begins its journey as fiberglass or carbon fiber, unrolled from massive bolts of cloth. This process stacks several layers of cloth (proprietary) laid out on vintage, wooden WWII parachute packing tables. Several patterns or panels are positioned on top of the stacked cloth. Using a rotary cutter, the worker then cuts out the various patterns, that are meticulously arranged to minimize waste.

Another worker takes the pre-cut cloth panels and arranges them in a prescribed pattern on a stylized human head form, and bushes the cloth with resin. This process "wets" the cloth and begins to bind the panels together in a random pattern, much like oriented strand board (OSB), which provides strength to the shell. Unfortunately, photography of this process was not authorized.

This sticky, wet assembly is removed from the head form and placed inverted into a steel two-piece mold (photo on the eft) that is used to form the helmet shell. This mold is mounted to a fixture that lowers a heavy duty bladder into the open void (top of mold), and inflated under high pressure to force the cloth and resin to the inside shape of the mold. The mold is also heated to 450-500 degrees Fahrenheit for several minutes until the resin is hardened. After the shell is removed, the edges require trimming and holes are drilled for attaching the chin and nape straps, oxygen mask receivers, visor housings, visor straps and for stitching the hand-sewn leather edge roll. The drilling process is mostly accomplished by water jet but, some are still hand drilled using a two piece jig with drill guides. These shells are produced one at a time.

The photo on the right is what the "raw" shells look like when they are removed from the molds. It should be noted that this photo is for

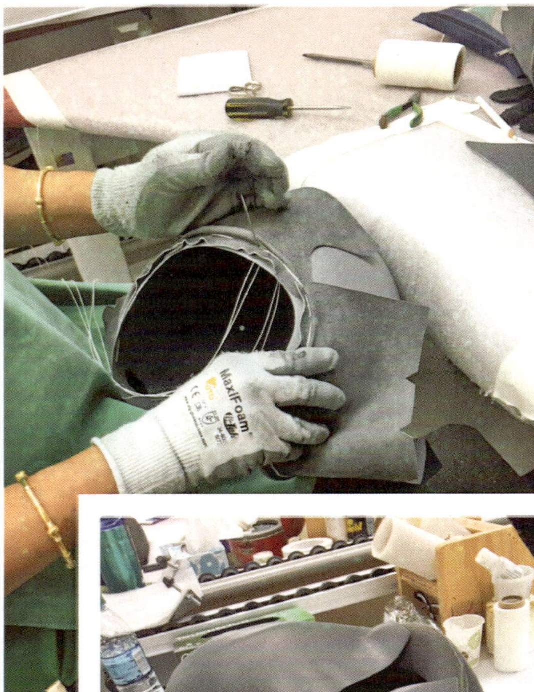

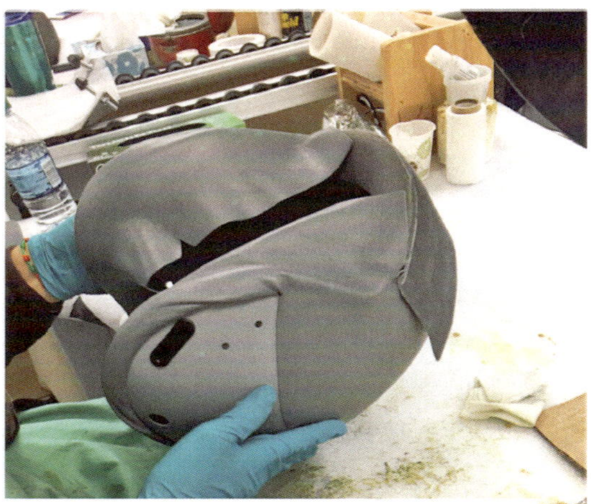

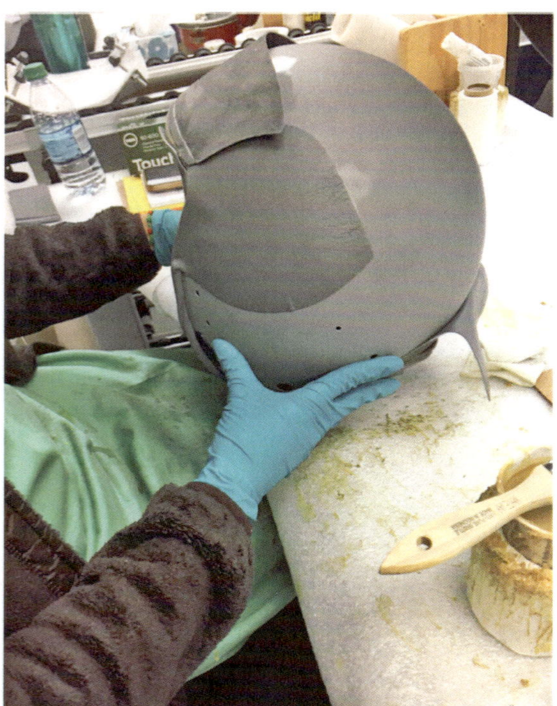

reference only. These are SPH-style shells and are destined for recycling due to production flaws deemed unacceptable in a highly regulated environment of very strict quality control inspections.

The shells are delivered to another small team of employees who are charged with filling minor imperfections, sanding, and priming the newly formed shells. The process is repeated until the surface is smooth and "signed off" for painting, which is one of only a few automated processes.

Once the shells are painted, they are once again inspected for quality control, and upon acceptance, get passed around like a new grandchild at a family reunion!

While I couldn't visit every station of the helmet assembly process, I did manage to spend some quality time at a station of particular personal interest to me; The hand sewn edge roll. Several years ago, I learned how to install a hand sewn leather edge rolls on flight helmets I was restoring as a hobbyist. It's not a terribly difficult skill, but it is tedious, takes practice, and it's difficult to do well!

I approached this station with respect, as this humble "student" was about to witness the "masters" at work. There were four women working with the efficiency of a well oiled machine and quietly conversing among themselves. Mark Jones introduced me, informed them that I was writing a book, and asked if they would be willing to discuss their respective duties and allow a few photos. I was also gently reminded that they had quotas, so get in, and get out!

I was well received into this inner fold, and I looked on in utter amazement as these women hand stitched the leather, glued the foam core to the shell and glued the leather to the inside of the shell. Any remnants of glue and pencil lines (outlining the "elephant Ears" and visor stops) were erased with a cloth dipped in methyl ethyl ketone (MEK) - a very effective solvent! These women were true professionals in every sense of the word, making their work look easy when I knew otherwise.

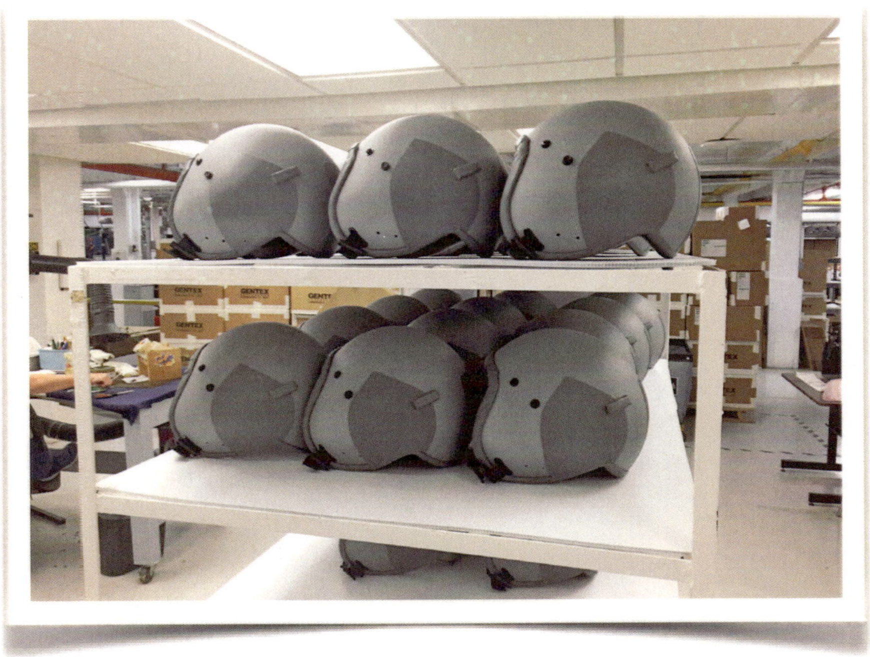

I was also impressed with the simple homemade jigs and fixtures that were created to make their work easier and more efficient. It is said that "necessity is the mother of invention" and it was evident at GENTEX. As a flight helmet collector and restoration hobbyist, I picked up on several tips and tricks to make my projects easier and more professional.

At this stage, the helmets are looking more and more finished but, much work remains to be done. These helmets will be passed along to many other work stations where one person will be responsible for one task until the build is complete. Integrated chin and nape straps are attached, comms are fitted, energy absorbing liners and Thermal Plastic Liners (TPL) are installed, oxygen mask receivers, visor assemblies, the list goes on...all hand installed...by individual employees...one helmet at a time. This is difficult to wrap your mind around considering GENTEX makes approximately 9,500 flight helmets annually.

This truly was a rare experience, and I feel privileged for the opportunity afforded me to witness a world class operation that continues to embrace the benefits and quality from hand crafted products.

With regard to the USAF Thunderbirds flight helmets, I was curious as to whether a "Plain Jane," "generic" grey HGU-55/P helmet from the GENTEX plant destined to become a USAF Thunderbirds flight helmet are produced differently. The short answer was, "Yes". Gibson & Barnes - an authorized GENTEX distributor, and custom flight helmet specialist to the Thunderbirds - order helmets from GENTEX with very specific requirements: HGU-55/P carbon fiber shells; White paint (not grey); NO edge roll; NO attachments. Gibson & Barnes receives the bare white shells, applies the red scallops (paint), stars, Thunderbirds graphics and finishes the "build" per Thunderbirds requirements. And, there it is!

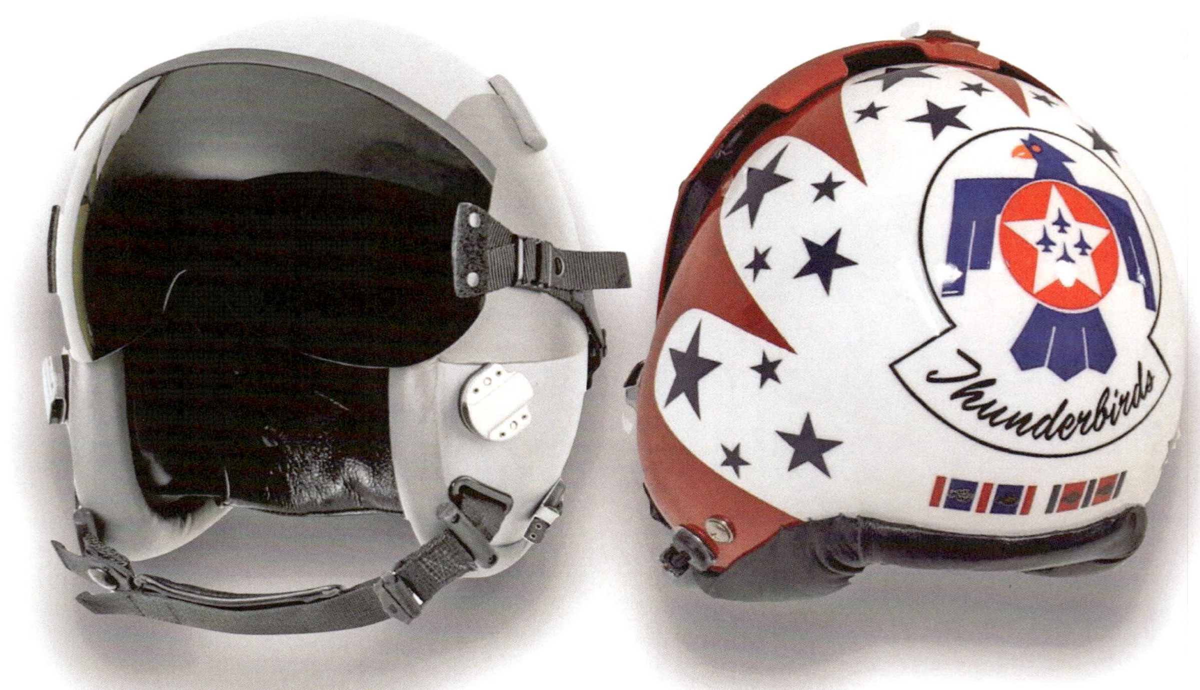

GENTEX

If you've got the right stuff, wear the right stuff.
When it comes to flying, aviation pros know it pays to own and use the best equipment money can buy.
In protective helmets, that means Gentex, the name synonymous with quality, innovation and service to pilots. Carefully designed, constructed and tested for the ultimate in comfort and safety, Gentex helmets help you fly your best no matter how challenging the conditions.
Available in off-the-shelf or custom-fitted and painted models. VISA or MasterCard accepted.
Gentex helps you show your stuff.
Gentex CORPORATION WESTERN OPERATIONS
2824 Metropolitan Place • Pomona, CA 91767 • (714) 596-6512
VISIT US IN THE OUTDOOR EXHIBIT AREA

GENTEX FLIGHT HELMETS

12
The Artists

What is an artist? Merriam-Webster's Dictionary[1] defines "artist" as follows:

artist

| *noun* | art - ist | \ 'är-tist \ **1** a. obsolete: one skilled or versed in learned arts; b. archaic: physician; c. archaic: artisan. **2** a. one who professes and practices an imaginative art; b. a person skilled in one of the fine arts; **3**. a skilled performer; especially: artiste; **4**. one who is adept at something, con artist, strikeout artist.

The Oxford English Dictionary[2] defines the older broad meanings of the term "artist":

1. A learned person or Master of Arts.
2. One who pursues a practical science, traditionally medicine, astrology, alchemy, chemistry.
3. A follower of a pursuit in which skill comes by study or practice.
4. A follower of a manual art, such as a mechanic.
5. One who makes their craft a fine art.
6. One who cultivates one of the fine arts – traditionally the arts presided over by the muses.

For the more contemporary, Wiktionary[3] defines the noun 'artist' as follows:

1. A person who creates art.
2. A person who makes and creates art as an occupation.

---

1 https://www.merriam-webster.com/dictionary/artist

2 http://www.oed.com/

3 https://www.wiktionary.org/

3. A person who is skilled at some activity.
4 A person whose trade or profession requires a knowledge of design, drawing, painting, etc.

Regardless of your preference, art is subjective, and like beauty, is in the eye of the beholder. In the eye of *this* beholder, the artists from Flight Suits, Ltd., now doing business as Gibson & Barnes personify the very meaning of the word. Is this subjective? Absolutely. Am I biased? Guilty.

Gibson & Barnes is a manufacturer of flight suits, uniforms, jackets, leather jackets, flying helmets and communications equipment. Customers are pilots, law enforcement officers, emergency medical personnel, military, test pilots, airline pilots and individuals.

Gibson & Barnes started manufacturing with two employees in a 1,000 square foot building in 1977. Over the years, slowly adding products and customers, they operate a 35,000 square foot facility and employ 115 people. Sales include products to more than 60,000 customers in the United States, Canada, and over 60 other countries.

Gibson & Barnes customers include the all branches of the U. S. armed forces, police helicopter units, hospital helicopter programs and NASA astronauts. They have also painted the US Navy Blue Angels and the USAF Thunderbirds flight helmets since 1984.[4]

4 Gibson & Barnes has also held the USAF Thunderbirds flight suit contract for the past 30 years.

In the Fall of 2018, Mark Jones with GENTEX introduced me to the Director of Gibson & Barnes, James A. Wegge, who, by the way, is a former Marine F-4 pilot. He graciously shared the company story with me. I learned that in 1984, Gibson & Barnes (formerly Flight Suits, Ltd.) purchased the custom-fit flight helmet department of Protection, Inc., which was owned by GENTEX since 1969. Protection, Inc. previously held the Thunderbirds flight helmet contract and that contract was transferred to Gibson & Barnes along with one employee. The first Thunderbirds flight helmets made by Gibson & Barnes were in 1984 for the 1985 show season.

A USAF Thunderbirds Flight Helmet as it goes through various stages of completion at Gibson & Barnes.

As a flight helmet painter and restoration hobbyist, I was not only an admirer of the high quality work product from the Gibson & Barnes artists, but a customer as well.

Mr. Wegge explained to me that GENTEX supplies HGU-55/P flight helmet shells to Gibson & Barnes pre-pained in white instead of the stock grey. The shells are void of any edge rolls, liners or external hardware to allow the artists to magically transform the plain white canvas into a Thunderbirds masterpiece! Following application of the graphics and red painted "scallops," the shell receives several layers of clear coat, wet sanding and power buffing until the finish is perfectly smooth and has a "wet" appearance. Only then, can a hand-sewn leather edge roll, liners and external hardware be installed.

There is a long-standing, ongoing debate in the flight helmet collector circles that the Thunderbirds (and Blue Angels) DO NOT use HGU-55/P shells. Turns out they do. You can take that to the bank. Hopefully, this should put this matter to rest.

In the mid-1990s, Flight Suits, Ltd. was using DuPont Imron polyurethane paint and clear coat, known throughout the aircraft industry for superior gloss, high quality and durability. Today, Gibson & Barnes has switched to PPG.

Unfortunately, due to highly restrictive California Environmental Protection Agency regulations, Gibson & Barnes moved their custom paint operations to Mexico. The Thunderbirds — and Blue Angels — flight helmets are sent to the Mexico plant for paint and graphics, and shipped back to the El Cajon plant for edge rolls and buildup.

The Thunderbirds typically place orders for new Team members in the late fall, and Gibson & Barnes customize the helmets in December and January. These are small orders — perhaps 4 helmets — so the window of opportunity to witness this process first-hand was not possible.

In the mid-1990s — the days of Flight Suits, Ltd. — I was fortunate to have the opportunity to work with some key personnel involved with Thunderbirds account. I received an eye-opening education on the detailed and meticulous process of transforming a basic white fiberglass shell into the iconic Thunderbirds flight helmet we know and love.

Over the years, I had developed a working knowledge of flight helmet restoration, painting and customization however, it was the application of graphics that captured my curiosity.

These images are authentic USAF Thunderbirds flight helmet graphics. The graphics (Thunderbird logo) are screen printed on clear acetate with an adhesive backing, while the stars are cut from high quality,

adhesive backed vinyl. Because the helmet shell is a complex compound curve — basically a sphere — the size of the two dimensional Thunderbird graphic is too large to conform to the curvature of the helmet shell. The solution is to cut the graphic into smaller, manageable components (top photo) that more easily follows the graceful curves of the shell.

Mr. Wegge explained, "Applying helmet art is a true multi-media process. It is a combination of printed film, machine-cut vinyl, hand-cut vinyl, hand-cut masking, and painting. It includes a considerable amount of surface preparation. The art is applied in stages and each helmet goes to the paint booth severa times. The process isn't simple, but the result is quite nice." Indeed.

The helmets are finished with multiple applications of clear coat followed by wet sanding and polishing. The process essentially "buries" the paint and graphics which appear to be under glass.

Because every Thunderbird helmet is hand-painted, by indiv duals and not machines or computers each one is ever so slightly different and, no two are identical, much like a fingerprint. Each one is as unique as the artist creating it, as well as the pilot wearing it; however, as Mr. Wegge points out, " The art on the helmet...will be slightly different on each helmet, but probably not detectible to the casual eye." In addition, very slight variations in the red paint will occasionally be encountered, explained by Mr. Wegge as changes in paint manufactures, and different formulas between U.S. and Mexican suppliers.

Following paint, graphics and clear coat, Gibson & Barnes installs a hand-sewn leather edge roll to the finished shell, a process we witnessed at the GENTEX plant. The visor housing and visor are installed along with the oxygen mask receivers, chin and nape straps and a liner. The new helmets are shipped to the Thunderbirds where Life Support Specialists inspect them and install communications.

In Chapter 11, Aircrew "Flight Equipment Specialist," we learned that in addition to the installation of communications, the visor housings are adorned with the USAF logo on the right side and the pilot's call sign and aircraft number on the left side. These graphics are applied over the top of the clear coat to allow a call sign or aircraft number to be changed, as several pilots rotate from one position in their first year, to a different position in their second year.

I'm sure you will agree, the Thunderbird artists are awesome; however, you may not know that the unique Thunderbird art work goes far beyond their custom flight helmets. In fact, it is all over their "office chairs!"

Earlier in this chapter, a cropped version of the photo above, featuring Major Jason "Jaws" Curtis'[5] flight helmet was included along with several others from the same period. The full image (above) reveals a custom painted head rest on the F-16 ACES II ejection seat, sporting an inverted "5" — the Lead Solo Pilot signature.

5 2013 - 2015 show seasons.

With the help of fellow flight gear collector Mike Baetslé of Belgium, I learned the identity of this artist, Jessica Esper, a designer, illustrator and painter doing business as The Graphix Chick[6] in New Mexico. Jessica tells me she became involved in painting for the Thunderbirds through a friend of hers, who was a former Team member. The rest is history!

The photos on this page are of Major Blaine "Deuce" Jones, solo pilot from 2012

to 2014. This head rest also features the classic inverted "5" for the Lead Solo position, followed by the letters "OLO" which looks like SOLO…Very clever!

[6] www.thegraphixchick.com

The images on this page are of Major Caroline "Blaze" Jensen, #3 Right Wing, 2012 - 2014, and her very inspirational head rest art.

The caricature art in the center of this work is a Walt Disney creation, "Fifinella," the incarnation of a mischievous imaginary gremlin, with magical powers. Pilots from WWII believed these gremlins played annoying pranks on them, and caused unexplained mechanical problems with their aircraft.

The Women's Air Service Pilots (WASP) of WWII fame, adopted "Fifinella" as their mascot, and the Disney Company granted them permission to use the image on patches, flight jackets and aircraft nose art.

"Fifinella" has stood the test of time and continues to inspire female pilots 75 years later!

Outstanding work, Jessica!

13
Nellis Air Force Base

In the late summer of 2018, I began giving consideration to attending the Ft. Worth Alliance Air Show featuring the USAF Thunderbirds *and* the Canadian Snowbirds! I was motivated by the outside chance of speaking to Thunderbirds Public Affairs about my book research. And, I had never witnessed a Snowbirds performance.

As I contemplated the 1,000 mile journey (one way), I reached out to Eric Schnaible, who was living and working in the Ft. Worth area. I was introduced to Eric by my brother,[1] who once worked for him while they were both serving in the U. S. Air Force. At the time, Eric was the International Communications Manager for the Lockheed Martin F-35 "Lightning II" program, and formerly Thunderbird #12 (Public Affairs) for the 1992 - 1994 show seasons. I had previously reached out to Eric seeking guidance on my book and he encouraged me to make the trip, suggesting I look him up when I arrived at the Alliance airport, as he may be able to provide me with a VIP Pass to the Thunderbirds tent! I was encouraged by this generous offer and immediately began planning for a road trip!

I arrived the Friday before the show weekend to absolutely miserable weather: low overcast, heavy intermittent rain and flood warnings! I was deeply disappointed. I was staying with a life-long friend[2] who shares a passion for aviation and lives only a few miles from Alliance airport. Despite the inclement weather, we could hear the distinctive sound of deep, throaty, turbine engines that could only emanate from military fighter jets. Intrigued, we drove to the airport to investigate. Much to our surprise, the Thunderbirds were performing a low practice show! Although encouraging, it wasn't long before the announcement was made by event organizers that the Saturday show would be cancelled due to heavy rain in the forecast. And, they

1 Major Scott A. Guidry, USAF (Retired)

2 Paul Stelly, Senior B-777 Captain; Master RC model builder / pilot; Machinist; Great friend.

were not wrong. It rained all day Saturday. The Sunday forecast was not looking much better, but we hoped for the best.

Sunday morning arrived with an overcast sky and a clear radar. The forecast called for rain later in the day. We checked the air show website and much to our surprise, didn't find any cancellation notices, so we headed out to the airport. Not long after arriving, I connected with Eric who made good on his promise of providing us with Thunderbirds VIP Passes! We watched the performance at show center among a small gathering of current and former Thunderbirds, family members and friends.

Following the show, Eric instructed me to meet him at a predetermined location where he would introduce me to Thunderbirds Public Affairs! I was greeted by Staff Sergeant Ned T. Johnston, NCOIC[3] of media operations. His responsibilities are to capture photos, to include aerial imagery, coordinate local- and national-level media flights, oversee social media accounts with more than 1 million followers and to represent the Air Force when interacting with communities around the world during the travel season.

SSgt. Johnston listened intently as I pitched my passion for the Thunderbirds, my research of the last 23 years and plans to publish a book. When he asked how he could be of assistance, I emphasized the importance of Thunderbirds support — without it,

[3] Non Commissioned Officer in Charge

the book would lack the validation necessary to make it a true, accurate and complete history.

I humbly asked for the opportunity to visit Nellis AFB, conduct interviews of both Public Affairs and Aircrew Flight Equipment personnel, and visit the Thunderbirds Museum for possible photos to be included in the book. SSgt. Johnston nonchalantly responded, "That seems simple enough." He advised that they were approaching the end of their show season, and my visit could be more easily accommodated during the winter training months between November and January. We exchanged contact information and SSgt. Johnston said he would be in touch with further details of a visit date. I thanked him, we shook hands, and he said, "See you at Nellis!"

"See you at Nellis!"

Staff Sergeant Ned. T. Johnston
USAF Thunderbirds
Public Affairs

It seemed too good to be true. Too simple. Too easy. Six months went by without a visit date. I was getting anxious, and to be honest, concerned the visit was only wishful thinking. But in the end, level heads prevailed. I was eventually informed that before a visit could be arranged, I needed to contact the Department of the Air Force, Public Affairs & National Media Engagement Office for formal approval and support. I wish I would have known about this months ago!

This U.S. Air Force Book Support Program provides authors a convenient, single location to submit requests for information, interviews, base tours and other related needs. A clearinghouse, if you will. OK, I'm a compliant person, so I submitted my request, along with a signed copy of my publisher's book contract, and once again... waited.

More glitches. My attached email files failed to open. Email messages were mysteriously relegated to my spam folder. I re-submitted hard copies of the required documents and mailed them, the old fashioned way. BINGO! The application was accepted, and I received an approval letter! This letter was forwarded to the Chief of Media Operations, Air Combat Command Public Affairs, who copied Thunderbirds Public Affairs for scheduling. After three more weeks of impatient waiting, the day had finally arrived. The wait was over; I was going to Nellis Air Force Base!

In mid-June, 2019, I received a long-awaited email from MSgt. Christine Powers, Superintendent, Public Affairs, USAF Thunderbirds! As the Superintendent, MSgt. Powers is responsible for managing media and community relations activities, and planning, coordinating, and scheduling all tours, orientation flights and media events for the Thunderbird team.

I was impressed, but not surprised that MSgt. Powers arranged everything. Thunderbirds tour. Pilot interviews. Visits to the Thunderbirds Museum and the Aircrew Flight Equipment Shop.

It took 8 months to get to this point. Email messages, letters, "pressing the flesh" at air shows across the U.S., and humbly complying with USAF public affairs protocol, but it was certainly worth the effort...I was now planning a trip of a lifetime!

57TH WING

First, a little history.

On May 1, 1950, Las Vegas Army Air Field was officially renamed Nellis Air Force Base, after the late 1st Lt. William Nellis, who was shot down while flying during the Battle of the Bulge.

Located on the Northeast side of Las Vegas, Nevada, Nellis AFB covers more than 14,000 acres, while the total land area occupied by Nellis and the restricted Nevada

Test and Training Range is more than 4,800 square miles. An additional 10,000 square miles of airspace north and east of the restricted range are also available for military flight operations.

Nellis is well known as the "Home of the Fighter Pilot" and "Home of the Thunderbirds," but that's just the beginning. Nellis also hosts "Red Flag[4]" and the U.S. Air Force Weapons School[5], the latter of which is part of the 57th Wing.

The 57th Wing — the most diverse wing in the Air Force — provides advanced, realistic, and multi-domain training focused on ensuring dominance through air, space, and cyberspace. The 57th Wing builds innovative leaders in tactics, training and high-end warfighting to ensure world-wide combat air forces are prepared for tomorrow's victories, while overseeing dynamic and challenging flight operations at Nellis Air Force Base, Nevada. The 57th Wing is comprised of the United States Air Force Weapons School, 57th Maintenance Group, 57th Operations Group, 57th Adversary Tactics Group, USAF Aerial Demonstration Squadron (Thunderbirds), USAF Advanced Maintenance and Munitions Operations School (AMMOS) and the 561st Joint Tactics Squadron[6].

Nellis, an integral part of the U.S. Air Force's Air Combat Command, is the pinnacle of advanced air combat aviation training. The base's all-encompassing mission is accomplished through an array of aircraft, including the A-10, F-15C/D, F-15E, F-22A, F-16C/D, HH-60 and most recently, the F-35A Lightning II[7].

---

[4] Red Flag is a two-week advanced aerial combat training exercise held 6 times a year to offer realistic air-combat training for military pilots and other flight crew members from the U.S., NATO and other allied countries.

[5] USAF Weapons School teaches graduate-level instructor courses, which provide advanced training in weapons and tactics employment to officers of the combat air forces.

[6] https://www.nellis.af.mil/About/Fact-Sheets/Display/Article/284155/57th-wing/

[7] https://www.mybaseguide.com/air_force/42-683/nellis_creech_afb_nellis_afb_history_and_overview

The Road Trip

According to Rand McNally, the distance from my home in Bagdad, Kentucky to Nellis AFB is about 1,900 miles. Yikes! I considered flying, but I needed to make several other stops along the way, some research related, some personal. Flying would also require me to rent a vehicle at each of my stops, so I decided to bite the bullet and embrace the cross-country road trip.

I started my journey 5 days ahead of my scheduled appointment with the USAF Thunderbirds. Yes, 5 days. This was by design, allowing me ample time to make the drive safely, and to allow for any unforeseen delays due to weather, road construction, or anything else Murphy's Law had to offer. While I had been to Las Vegas many times during my life, this would be my first trip to Nellis! I felt like a kid at the beginning of summer vacation!

DEPARTMENT OF THE AIR FORCE · UNITED STATES OF AMERICA · MCMXLVII

AIR COMBAT COMMAND

USAF WARFARE CENTER

Nellis Air Force Base

Just prior to my departure for Nellis AFB, I received an email from MSgt. Powers asking me if I could spend a second day with the Thunderbirds, as it would be helpful for them to fulfill all of my requests. I couldn't type "YES" and hit the send key on the computer fast enough!

It also occurred to me that several months earlier, I reached out to, and requested an interview with Lt. Col. Caroline "Blaze" Jensen, who flew the #3 position for the Thunderbirds from 2012 to 2014. After not receiving a response, I was considering this a strike out. Not so fast. I received an email from the 926th Wing Public Affairs on behalf of Lt. Col. Jensen advising me that she was receptive to an interview! It gets better. The 926th Wing is at Nellis AFB! And, she was willing to be interviewed when I was at Nellis with the Thunderbirds! This journey just kept getting better!

I arrived in Las Vegas, NV on the afternoon of Monday, 01 July, the day before my first scheduled appointment. It was 105 degrees Fahrenheit. I was tired from the long days on the road and hard miles on my aging body. My muscles ached, and my eyes were strained, but the physical wear and tear couldn't come close to diminishing my mental excitement. *I was going to visit the Thunderbirds!*

Once I got settled into my hotel room, I established contact with MSgt. Powers. Plans were already changing and my visit would now be compressed into one day. Not to worry, nothing had been cancelled. I received the following itinerary:

0700 hours: Meet MSgt. Powers at the Nellis AFB Visitor Center, with 2 forms of photo identification.
0730 hours: Interviews with Thunderbird #2, Major William "Boar" Graeff and Thunderbird #6, Captain Michelle "Mace" Curran[8] in the VIP Room of the Thunderbird Museum.
0800 hours: Interviews of Life Support Specialists and photographs of the Aircrew Flight Equipment Shop.
1000 hours: Thunderbird Museum and Hangar tours.
TBD (p.m.) Interview of Lt. Col. Caroline "Blaze" Jensen, Thunderbird #3, 2012-2014 show seasons, in the VIP Room of the Thunderbird Museum.

I was ready for this. Route plan: Check. Two alarm clocks set: Check. Starbucks: Check.

Tuesday morning had finally arrived. My instructions from MSgt. Powers were clear and concise. It was show time!

I arrived at the Nellis AFB Visitor Center 30 minutes early for a background check and vehicle pass. I was greeted by the stunning Thunderbird diamond monument on the right.

MSgt. Powers arrived, sponsored my visitor pass and we proceeded to the Thunderbird hangar. Driving onto Nellis AFB and approaching the Thunderbird hangar was like a

[8] Captain Curran promoted to Major the day prior to this interview.

mirage — the only white structure nestled in a sea of desert sand colored buildings. This was surreal.

United States Air Force Thunderbirds

I had seen many photos of the "Home of the Thunderbirds" but being here

and witnessing this first-hand was truly exciting and will provide me with a memory that will remain vivid for the remainder of my life. The grounds are immaculate. Impressed, but not surprised.

Just inside the main (museum) entrance is a large, 3 dimensional Thunderbird emblem hanging proudly on the wall, leaving little doubt that you have entered a very special space.

The emblem was next to the entrance to the Thunderbird VIP room (above), where MSgt. Powers set me up for my scheduled pilot interviews. The table is typically much more orderly, as you might expect, but the "mess" is a result of me "moving in" for the day.

The glass table top was etched with the Thunderbird emblem in the center. Each of the Thunderbird officers' number inside of a circle of 12 stars was etched in the glass in front of each of the 12 seats.

The VIP Room is adjacent to the Thunderbird Museum, which offers the visitor a very diverse selection of memorabilia and artifacts from the Team's inception in 1953 to present day. I have selected only a sample of the exhibit items. All photos on this page were taken by the Author.

All photos on this page are courtesy of the U S A F Thunderbirds Museum.

Beyond the Thunderbird Museum is the hangar. Words simply can't describe the floor — clean and spotless are understatements! The top photo is courtesy of the USAF Thunderbirds, and the inset is by the author. When asked if I could photograph the Thunderbird emblem, I was told "Yes"; however, I was prohibited from touching or standing directly on the emblem, I could not touch the stanchion or rope, nor could I stand on any shadows cast by the stanchion and rope. Roger that.

Just to the left, and out of view of the F-16 in the photo on the previous page, is an unassuming door with a sign above that simply reads, "The Perch." This very exclusive "club" is a place where Team members can relax, eat lunch, or enjoy an adult beverage at the end of the day.

In recent years, this has also become the gathering space of choice for ceremoniously awarding new pilots their "Red Helmets."

The photo of the Thunderbird flight helmet beer tap is courtesy of Mike Moore. Unfortunately, this one-of-a-kind item no longer graces the bar.

Inside the hangar — hanging on the walls — are 3 stunningly beautiful models of the most recent aircraft flown by the Thunderbirds: The F-16 "Fighting Falcon," the T-38 "Talon" and the F-4 "Phantom." According to MSgt. Powers, whenever a new Commander / Leader takes over the Team, his name is placed on the canopy rail of the model, just as it is on the real aircraft.

Behind the massive flag draped in the hangar is another door that leads to a host of support teams, offices and workspaces, including the Aircrew Flight Equipment (AFE) Shop. The AFE shop tour and interviews was one of the main reasons for this visit so it only makes sense that AFE deserves a dedicated chapter...read on!

14
Aircrew Flight Equipment Specialist

Aircrew Flight Equipment Specialists manage, performs and schedules inspections, maintenance and adjustments of assigned aircrew flight equipment (AFE), aircrew chemical defense equipment (ACDE), associated supplies, and inventories assets.

Duties performed by these specialists include:

Inspects, maintains, packs and adjusts aircrew flight equipment such as flight helmets, oxygen masks, parachutes, flotation devices, survival kits, helmet mounted devices, aircrew night vision and other ocular systems, anti-G garments, aircrew eye and respiratory protective equipment, chemical biological protective oxygen masks and coveralls, and other types of AFE and aircrew chemical defense systems. Repairs fabric and rubber components, including protective clothing, thermal radiation barriers, flotation equipment, and various parachutes.

Evaluates problems and determines the feasibility of repair or replacement related to inspecting and repairing fabric, rubber equipment and parachutes. Evaluates work orders for fabrication of authorized items.

Installs and removes aircraft-installed AFE. Uses various types of test equipment such as altimeters, oxygen testers, leakage testers, radio testers and other types of testers to conduct reliability testing on AFE and ACDE. Maintains inspection and accountability documentation on AFE issued to aircrews or prepositioned on aircraft.[1]

[1] https://www.thebalancecareers.com/1p0x1-aircrew-equipment-air-force-job-descriptions-3344267

United States Air Force Thunderbirds Flight Helmets

Pilots, as well as all flight crew members, must be knowledgeable about safety equipment and survival procedures in the event that a mission does not proceed as planned, and in a worst case scenario, their aircraft is damaged, on fire, or incapable of maintaining flight.

All USAF flying squadrons employ Aircrew Flight Equipment (AFE) Specialists[2], and the Thunderbirds are no exception. Before and after any practice flights, aerial demonstrations or flight of any description, an AFE Specialist has inspected and insured that all life support equipment is serviced and functioning properly. This includes, but is not limited to flight helmets, oxygen masks, anti-g suits and parachute harnesses. The pilots implicitly trust the AFEs with the critical role they play in the interest of their safety.

Tech. Sgt. Edwin Portan (Bottom Left) and Tech. Sgt. Amber Alumpe (Top Right)
USAF Thunderbirds Aircrew Flight Equipment.

[2] Formerly two specialties, Life Support, and Survival Equipment.

The top image is not Photoshopped! This is a rare photo opportunity taken during what used to be an annual event where the USAF Thunderbirds and USN Blue Angels would alternate visits to NAS Pensacola and Nellis AFB to share best practices, perform shows for each other, and have a chance to fly as an "observer" with the opposing team.

I was fortunate enough to witness two of these events, one in 1998 at NAS Pensacola, and the second in 2019 at Joint Base Andrews.

Left, Staff Sgt. Jose Ibarra conducting a 30-day oxygen mask and flight helmet inspection. (U.S. Air Force photo by Staff Sgt. Kristina Overton)

A close examination of the flight helmet pictures on this page reveals the stars are slightly less obtuse (left image below) from the "standard" pentagram, or regular star polygon (right image) typically seen on most examples:

My introduction to the Thunderbirds Aircrew Flight Equipment (AFE) Shop (formerly Life Support) dates back to 1996 when I made that first "cold" call and spoke with Staff Sergeant Jeff Kessler.[3] We shared a common first name as well as a passion for the USAF Thunderbirds. Over the course of about a year, I spoke with SSgt. Kessler many times by phone, and arranged to meet him in person when the Thunderbirds were performing an air show in my hometown, Lafayette, LA.

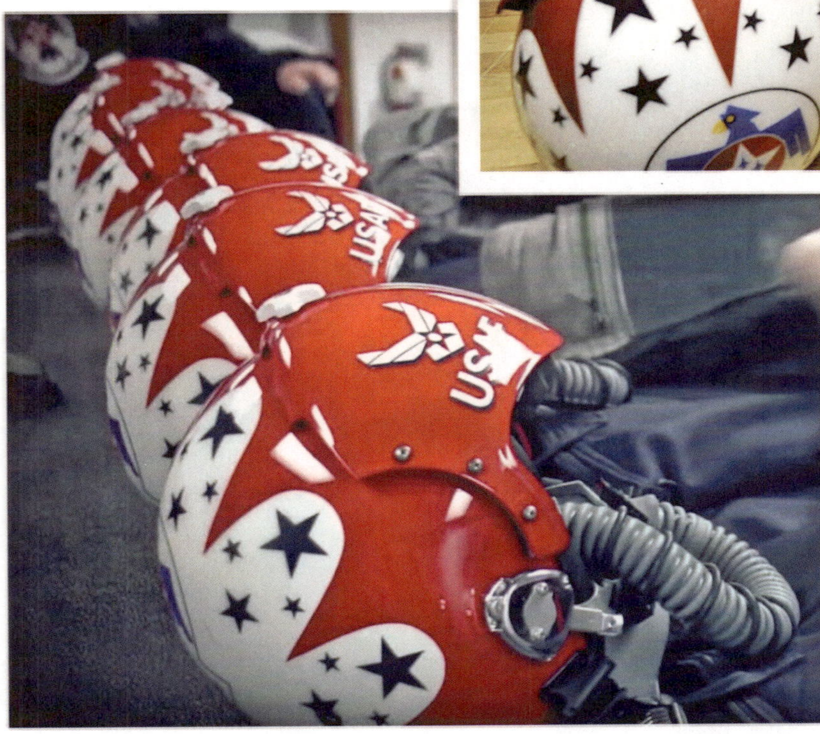

I was afforded the opportunity to study, examine and photograph an authentic Thunderbirds flight helmet, which marked the beginning of my research for this book.

A copy of the air show program from 1996 is shown on the next page. I've been referred to as a "pack rat" among other things!

[3] Currently Master Sergeant, USAF (Retired)

SSgt. Kessler was my first of many Thunderbird[4] contacts over the years and his kindness, patience and generosity have made a lifelong impression on me. Without his interest and willingness to assist me, this project may not have been possible. I am truly grateful.

Little did I know at the time that 23 years later, I would still be conducting research, interviewing Thunderbird team members and traveling thousands of miles to make sure this story is as complete and accurate as possible. Nor did I think I would ever have the chance to travel to Nellis AFB, "Home of the Fighter Pilot" and USAF Thunderbirds! That was about to change!

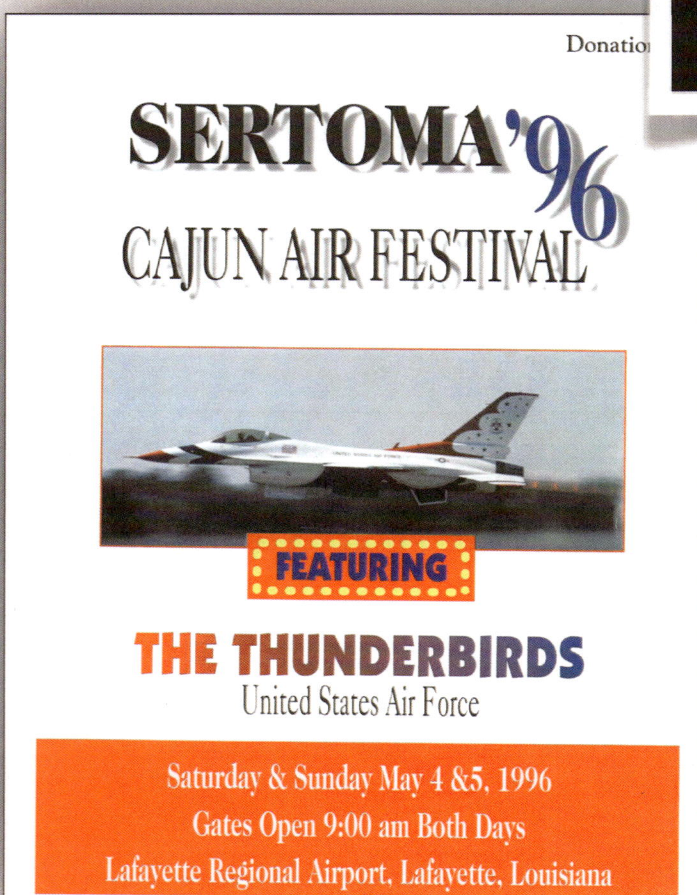

[4] SSgt. Dereak Sikes, Crew Chief #2; SRA Warren Hensman, Aircrew Life Support

New Orleans Joint Reserve Base, Louisiana 1997
VIP pass compliments of SRA Warren Hensman
Aircrew Life Support

With the generous assistance from the Thunderbirds Alumni Association[5] (TBAA), I had the opportunity to interview a few "vintage" Thunderbirds Life Support Specialists — one from the 1965 to 1968 show seasons (F-100 "Super Sabre"), one from the 1979 to 1981 show seasons (T-38 "Talon") and the third from the 1984 to 1987 show seasons (F-16 "Falcon"). Since all three members were located several states away from me, we agreed to share information through telephone interviews and email exchanges. Here are their fascinating stories.

SSgt. James "JB" Berry
USAF Thunderbirds Life Support
1965 to 1968

I consider myself very fortunate for the opportunity to interact, albeit by telephone, with so many people who played such critical roles in support of the USAF Thunderbirds. SSgt. James "JB" Berry is one such person.

When I spoke to "JB" one thing was certain from the onset of our conversation — he remains passionate about his service with the Thunderbirds despite the passing of more than 50 years! I was touched and impressed with his zeal and attention to detail.

I was surprised when "JB" told me he was the only Life Support Specialist on the team when he served — difficult to fathom in this day and time, given the magnitude of their responsibilities. This was the era of the F-100D "Super Sabre" and "JB" has fond memories of VIP and celebrity orientation flights, the most notable of which was Frank Sinatra, Jr. He said he was responsible for briefing the passengers on emergency and ejection procedures in the event the flight didn't proceed as planned. Safety First!

---

[5] Colonel Michael "MO" Beale, USAF (Retired); TBAA Chairman

When I asked "JB" about flight helmets, he remembers working with a company out of southern California that was contracted to provide the custom painting. I asked him if it was Protection, Incorporated out of Pomona, CA, and he said, "That sounds familiar." He said a "new style" of helmet came out around 1961, and the Team began using them. He directed me to a copy of a Thunderbirds Alumni Association publication, *ThunderRolls*, that featured a photo of "Wild Bill" Fenton (right) wearing one of the new helmets. He remembers the helmets had an improved oxygen mask retention system.

The helmet in this photo (right) is a USAF P-3, which was used by the Thunderbirds during the F-84F era. The Thunderbirds upgraded their helmets about the same time as they transitioned to the F-100; however, as mentioned in Chapter 4, they chose the TOPTEX over the P-4. The pictures below compare and contrast the P-3, P-4 and the TOPTEX, which a-e very similar, and remain confusing, even to seasoned collectors and historians.

The helmet on the left is a P-3, the middle example is a P-4 and the one on the right is a TOPTEX. The P-3 and P-4 originally used leather tabs with snaps for attaching the oxygen mask to the helmet; however, the TOPTEX used nickel plated, Hardman oxygen mask bayonet receivers, which is what I believe "JB" was referring to when describing the "new" helmets.

Capt. Tony McPeak
(Photo credit: USAF Thunderbirds)

With excitement in his voice, "JB" shared the story of when Capt. Tony McPeak[6] ripped the wings off an F-100 during an air show performance. The location was Laughlin AFB, just outside of Del Rio, TX. The maneuver was the the "Bomb Burst."

Capt. McPeak wrote of the incident, "I start the aggressive pull into the vertical. And the aircraft explodes." McPeak is forced to eject when the wings are ripped from the aircraft[7] and bursts into flames. McPeak continues, "My helmet is missing. Where did it go? I look up and see a couple of chute panels are torn, several shroud lines broken, and there's one large rip in the crown of the canopy." The first Thunderbird accident...in front of an air show crowd!

According to Capt. McPeak, "When I came out of the jet, the near sonic wind blast caught my helmet, rotated it 90 degrees and ripped it off my head. It was found on the ground with the visor down, oxygen mask still hooked up and chin strap still fastened."

[6] General McPeak served as the 14th Chief of Staff of the Air Force from 1990 to 1994.

[7] The wing attach points were weakened by stress fractures.

"I didn't get enough separation from the seat which somehow contacted my parachute canopy causing the large tear." Capt. McPeak miraculously walked away from the accident.

"I start the aggressive pull into the vertical. And the aircraft explodes."

Capt. Tony McPeak
USAF Thunderbirds

I asked "JB" if he witnessed the "incident" and he said, "No, I was in the hangar!" He said he was able to examine Capt. McPeaks parachute after the ejection, and described the hole in the canopy as, "Large enough to drive my pickup truck through!" I had to ask, "Did you pack McPeak's parachute?" "JB" replied, "Oh God, no, that was the parachute shop!"

"JB" separated from the Air Force following his tour with the Thunderbirds and continued providing life support services to the U. S. astronauts in the NASA manned space program. Thank you for your service, Sir!

Photo credits: https:generalmcpeak.com/photos/

Sgt. Mike Turkington[8]
USAF Thunderbirds Life Support
1979 to 1981

I spoke with Sgt. Mike Turkington in January of 2019 and learned he has enjoyed a life-long passion for the USAF Thunderbirds and he has fond memories of attending air shows at a young age. He told me that during his service with the Team — like"JB" in the previous story — was the only Life Support Specialist for the first 18 months, requiring him to travel with the Team to many air shows! Of course, this was in the back seat of a T-38A "Talon" which is not a bad way to travel! As a result, Mike said he accumulated about 100+ hours of flight time! Admittedly, I'm a little envious!

The T-38 did not have refueling capability, so on long cross country flights to and from Nellis AFB and the show sites, they were occasionally required to make a fuel stop.

[8] Staff Sergeant at separation from the Air Force

On one particular flight, Lt. Col. David L. Smith[9], Commander / Leader, was having a problem with a "sticking" oxygen mask valve. Mike said the MBU-5/P oxygen mask had a nasty reputation for this problem. He said it required the wearer to exhale sharply — sometimes repeatedly — to clear the valve. It was reportedly aggravating and, at times, made communications difficult.

Lt. Col. David L. Smith
Protection, Incorporated HGU-26/P Flight Helmet
(Photo credit: USAF Thunderbirds)

Mike said the "Boss" had about enough of his sticking mask valve, so when they got on the ground for a scheduled fuel stop, he strongly "summoned" for Turkington! The "Boss" removed his mask, handed it to Mike and said, "Here...give me your mask...now it's your problem!"

Mike said the Thunderbirds were still working with Protection, Incorporated (PI) at the time of his service with the Team. PI was a custom flight helmet producer that painted these beautiful Thunderbirds flight helmets. It was interesting to learn that the Life Support Shop would send PI an HGU-26/P flight helmet from USAF inventory to be custom painted. PI could have supplied the shells but, they came at a high cost.

[9] Killed on 08 September 1981 when his T-38 ingested a flock of seagulls, causing him to eject. The low altitude ejection did not allow enough time for his parachute to fully deploy, killing him on impact with the ground.

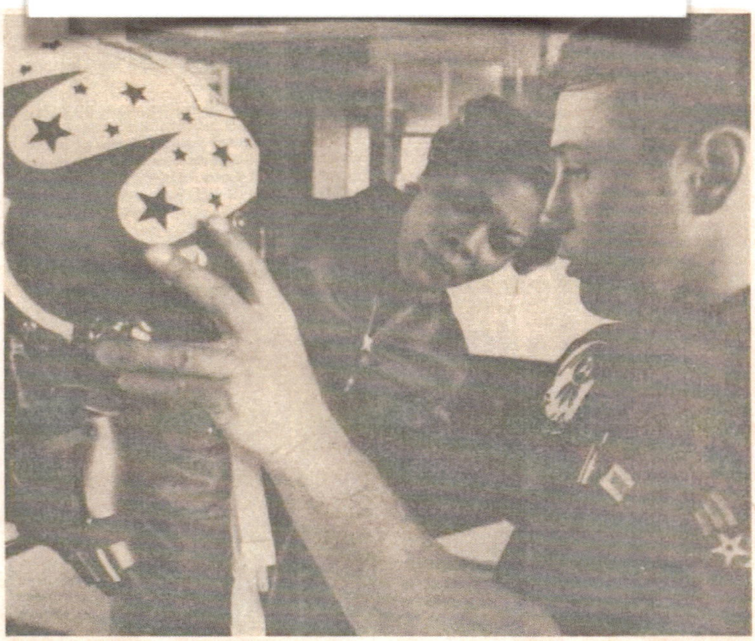

WARWICK WINGS IT — Singer Dionne Warwick checks over a helmet and oxygen mask with Air Force Sgt. Mike M. Turkington, Nellis AFB life support specialist. She was at the base Tuesday for an orientation flight in one of the U.S. Air Force Thunderbirds' T-38 Talon Jets. The Thunderbirds are the reknowned aerial demonstration team.

PI would make periodic "sales calls" to the Life Support Shop about twice a year to "press the flesh" and insure everyone was satisfied with their work. On one occasion, a custom Thunderbirds flight helmet was delivered that was described as "slightly off color" and rejected by the team. Mike said he was able to purchase the helmet for a "fair" price for his use on the team.

Mike — like all Life Support Specialists — is responsible for briefing those lucky passengers chosen for VIP and celebrity orientation flights. Mike had the good fortune of working with Country Music recording artist and icon, Barbara Mandrell! In the photo on the left, Sgt. Turkington briefs Barbara Mandrell on the proper operation and use of the HGU-26/P flight helmet prior to her VIP flight — lucky guy!

The photo and newspaper article (left) show Sgt. Turkington briefing the

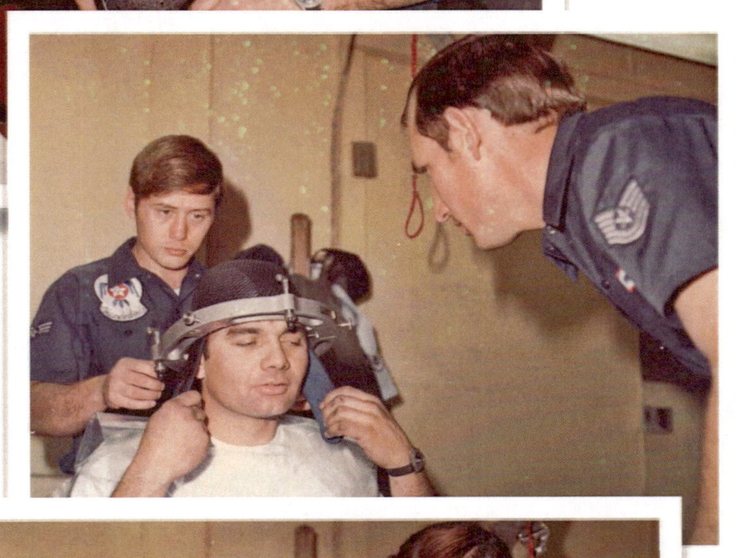

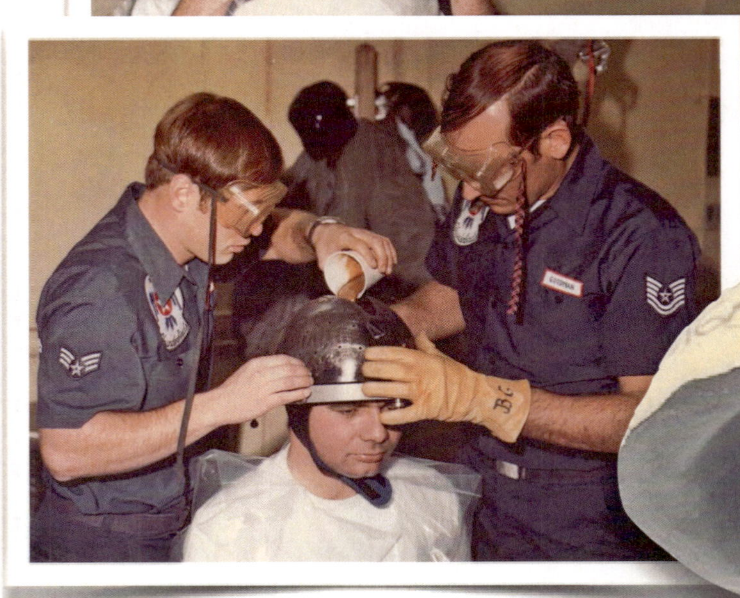

legendary Dionne Warwick on the flight helmet she would be wearing on her orientation flight. The briefings included familiarization with parachutes, oxygen masks, aircraft cockpit and ejection seat procedures.

Sgt. Turkington also shared a series of very interesting photos that illustrate the process a new pilot goes through when being fitted for a custom flight helmet liner.

The photos on the left are of Sgt. Turkington and TSgt. Jim Goodman making a form fitted liner for a flight helmet. The process involves placing a mold / form on the wearer's head, and filling it with a two-part, expanding foam product. When dry, the mold is released, and the foam form covered with leather.

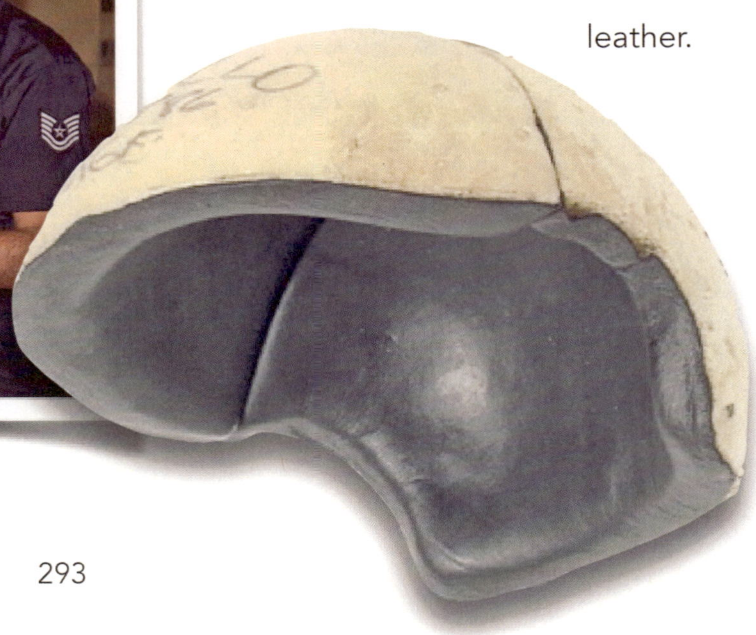

The photo on the bottom of the previous page is an example of the two-piece, custom, form fitting liner produced from the molding process. The tan colored portion on the top of the image is the custom formed portion of the impact absorbent foam, covered in black leather. These custom fit liners were far superior to the leather-faced pads that were supplied with standard issue flight helmets. These liners were more comfortable, and eliminated what pilots called "hot spots" on the head caused by the pads that came standard with USAF government issue flight helmets.

The photos on this page are of the personal helmet belonging to Sgt. Mike "Turk" Turkington. The "off" color he referred to can be seen in the photo on the right. The white "field" inside of the Thunderbird emblem appears slightly yellowed. Still beautiful, nonetheless.

I had a good laugh when Mike told me that his helmet was a size medium, and was occasionally used by female passengers, such as Barbara Mandrell and Dionne Warwick. The Team's spare helmets were all size large, and Mike's provided a better fit. It's a cool story, but Mike said after the flights, it took him about an hour to clean all of the makeup from the interior of the helmet and oxygen mask! Well worth the memories, I'm sure!

SSgt. Jeffrey Schenkelberg[10]
USAF Thunderbirds Life Support
1984 to 1987

SSgt. Schenkelberg (kneeling) checking the parachute of an unknown Sergeant prior to an incentive ride.

As we move forward in this Life Support journey, we have made our way to the very early years of the F-16 "Falcon" era. Unlike SSgt. Berry and SSgt. Turkington in the previous two stories, SSgt. Schenkelberg had a small team of Life Support Specialists to help distribute the work load. Despite the larger team, work days could stil run 12 to 16 hours during show season.

SSgt. Schenkelberg — like his predecessors — had fond memories of the VIP and orientation flights. As an interesting piece of trivia, the Thunderbirds were still using the T-38 in 1983 and 1984 for their #8 Advance aircraft as well as media / orientation rides despite having transitioned to the

[10] Master Sergeant Jeffrey Schenkelberg, USAF (Retired)

F-16. Two such flights that came to mind during our conversation, were Wayne Newton and Secretary of the Air Force, Edward C. "Pete" Aldridge Jr.[1], the latter of which had him a little "rattled" from the pressure to get the job right!

Regarding his work with flight helmets, Schenkelberg told me the Life Support Shop was working with Flight Suits, Ltd. (FSL) of El Cajon, CA during his service with the Team. He recalls ordering helmets from FSL, which they would paint, apply graphics, clear coat, and ship to the Thunderbirds. The shells were were then built up by Life Support, adding the visor housing and visors, custom liners, communications, chin and nape straps and oxygen mask receivers.

On some helmets, FSL would install custom leather, hand-sewn edge rolls, but not always. Schenkelberg says he became quite skilled at installing the edge rolls and was excited when the SCOTT "snap-on" edge rolls became an available option that made the job easier and faster.

These SCOTT "snap-on" edge rolls came in various sizes to correlate to the size of the helmet shell. This option eliminated the need to cut out leather patterns, hand-sew the leather to the helmet shell, glue the foam to the shell, then carefully stretch the leather over the foam and glue it to the inside of the helmet shell.

The SCOTT "snap-on" assembly kit was a plug-and-play item molded from a soft polyester foam and configured to fit over the helmet shell contour and provide a

[1] In office April 6, 1986 – December 16, 1988

complete seal around the wearer's face and neck. Secure attachment to the helmet shell is provided by two plastic inserts that afford a firm grip over the shell surface. The edge roll is enclosed by a leather cover tailored to fit smoothly over the foam core after the essential fitting operations are performed. Edge roll assemblies are provided in three sizes: Medium, Large and Extra Large.

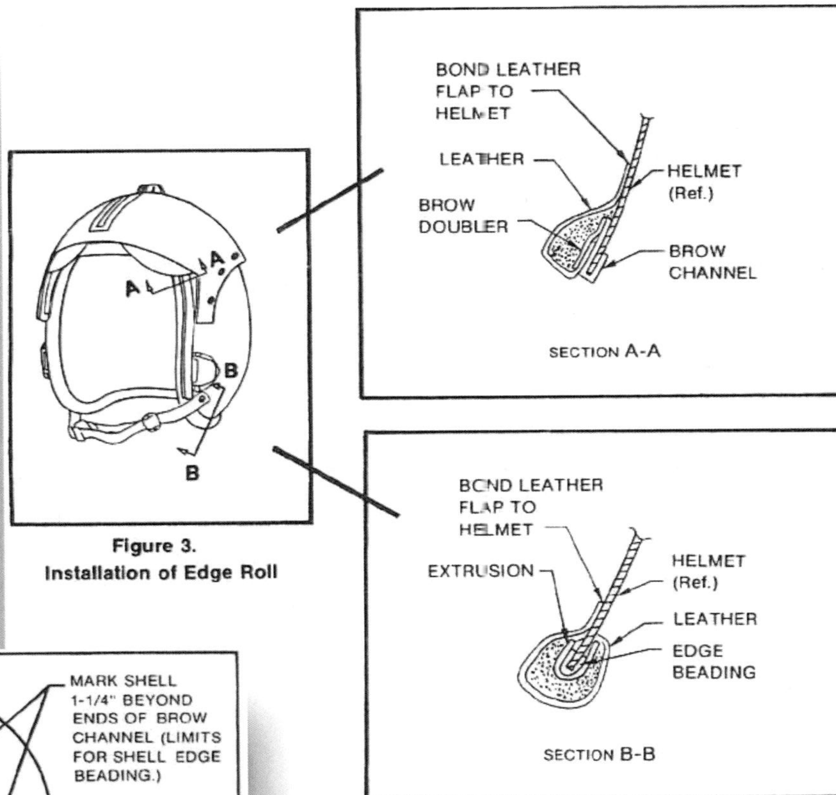

Figure 3.
Installation of Edge Roll

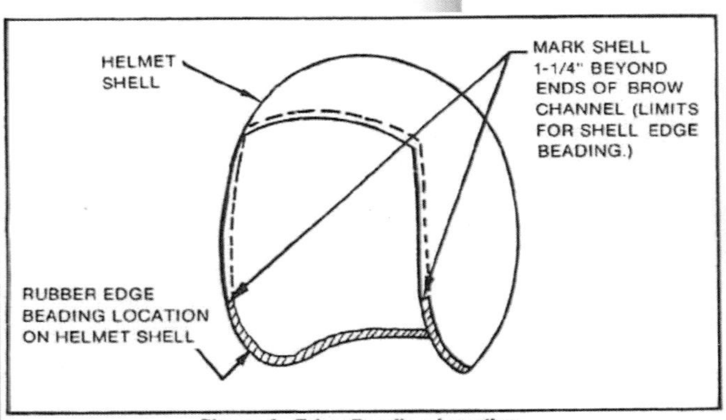

Figure 2. Edge Beading Location

The use of snap-on edge rolls by the Thunderbirds surprised me...still searching for that elusive example!

On this page are some excerpts from the installation instructions, which are simple and straightforward.

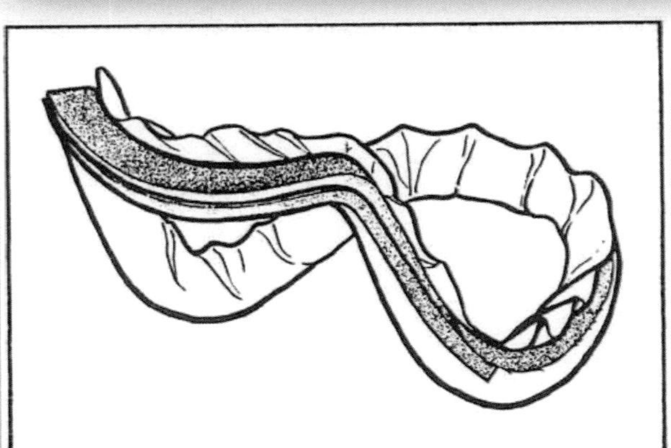

Figure 1. Edge Roll Assembly

Unlike a hand sewn leather edge roll where the shell is stripped of all attachments and assemblies, only the chin and nape straps, and ear cup assemblies need to be removed for attachment of the snap-on edge roll assembly.

The oxygen mask receiver hardware, helmet liner sections and visor assembly should not interfere with edge roll installation procedures, with the exception of the sufficient loosening of the lower visor mounting screws to allow the edge roll brow channel to slide into place under the visor tracks. The helmet liner sections are removed for the final installation operation which involves bonding the leather edge roll cover to the interior of the helmets shell.

SSgt. Schenkelberg also said that when he first joined the Thunderbirds, they had just switched from the HGU-26/P flight helmet to the HGU-55/P. The HGU-55/P flight helmets were equipped with lightweight oxygen mask bayonet receivers, which are not very aesthetically pleasing (photo on right); however, he was pleasantly surprised when the team kept the chrome plated, cast aluminum receivers used from the late F-4 and throughout the T-38 eras.

The Thunderbirds were previously sourcing the chrome oxygen mask receivers from Protection, Inc. and FSL until they learned that there was a shop on base that had chrome plating capability. The oxygen mask receivers were subsequently plated on-site, saving time and money.

Aircrew Flight Equipment Shop
Nellis AFB

Because of my flight gear collecting passion, and fascination with how things are made, I fancy myself a life support "hobbyist" or "enthusiast" by default. Despite no formal training in this critical discipline, I developed some useful skills from years of disassembling and re-assembling countless flight helmets and oxygen masks with the help from a modest library of USAF technical manuals on aviation life support equipment and flight helmet maintenance.

This humble student now stood in awe and amazement on the threshold of the mother of all life support shops. A moment of silence, please.

MSgt. Christine Powers, Superintendent, Public Affairs, arranged for me to meet with MSgt. Edwin Portan, AFE Flight Chief. MSgt. Portan manages a team of Aircrew Flight Equipment Specialists that include TSgt. Jasper Roberts, AFE Non Commissioned Officer in Charge, SSgt. Kyle Boddie, AFE Shift Lead, and SSgt. Joshua Metzger, AFE Craftsman, all pictured below in the AFE Shop.

From left to right, SSgt. Joshus Metzger, SSgt. Kyle Boddie, MSgt. Edwin Portan and TSgt. Jasper Roberts.

The AFE Shop was as I expected: Neat and organized, reflecting the pride of every USAF Thunderbird. To say I was impressed would be an understatement.

There was no doubt I was in the AFE Shop, from the large AFE disk on the door, to the fiery red Thunderbird emblem on the end of the work bench, to the visor housings personalized with the names of the AFE Specialists. These decals, or stickers, are referred to as "Gerbers" which comes from the product name. The graphics are a 3M product called Scotchcal 220, custom formulated for Gerber Scientific Products. Now you know.

I learned that flight helmet orders placed with Gibson & Barnes are based on incoming pilots' helmet size, Medium, Large or Extra Large. As the budget allows, they sometimes maintain a few "stock" helmets in their limited inventory. These helmets arrive freshly painted with a hand-sewn leather edge roll. The AFE Shop adds the appropriate "Gerbers" and communications, including soft, combo ear seals with noise reduction kit sourced from Pro Flight Gear, Aviation Life Support Equipment, as well as a Zeta Liner from Oregon Aero.

The two photos on the left are of a Thunderbird flight helmet as it is received from Gibson & Barnes. The red, white and blue skull caps have also been discontinued, as it was a custom order and an unnecessary expense. The Team currently uses solid red skull caps, commercially available through Gibson & Barnes.

Life Support plays a critical role regardless of the mission. Thunderbirds AFE are the best of the best.

Immediately following each sortie — whether training, or an air show performance — every flight helmet and oxygen mask are cleaned and inspected to insure they perform as intended.

On the day of my visit, the flight helmets, helmet bags, oxygen masks and anti-g suits were neatly lined up on the AFE bench, clean, inspected and ready for the crew chiefs to pick up the equipment for the pilot and respective jet for which they were responsible.

Following a morning launch and practice over the Nevada desert, the flight equipment was returned to the AFE Shop by the Crew Chiefs, and once again, cleaned, inspected and lined back up prior to an afternoon launch. The pilots place a great deal of trust in the AFE Specialists. There is no room for error. This is serious business.

The photo on the left (author's photo) shows the flight equipment lineup prior to a scheduled launch. The photo on the right (author's photo) is of SSgt. Kyle Boddie, AFE Shift Lead as he meticulously cleans and inspects a flight helmet and oxygen mask prior to a practice flight.

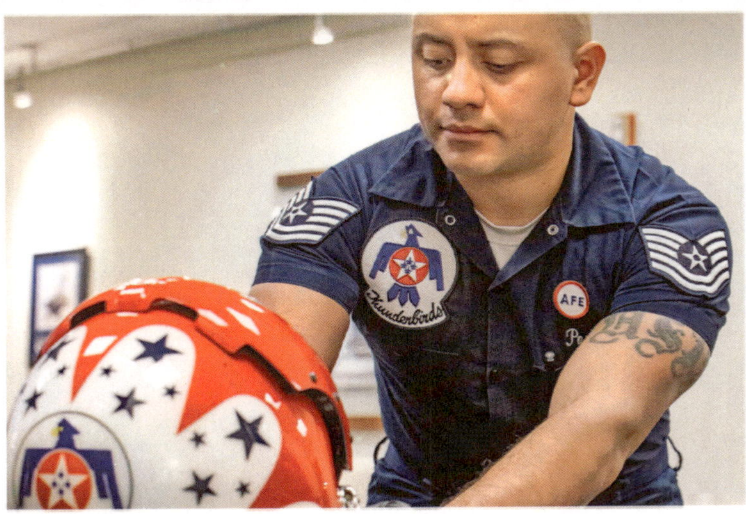

On this page, we see MSgt. Edwin Portan (then Technical Sergeant) carefully cleaning and inspecting a flight helmet and oxygen mask at a show site (USAF Thunderbirds photos).

I capitalized on the opportunity to further discuss the Combat Edge modifications made following the April 2018 crash as discussed in Chapter 7 - Combat Edge Flight Helmet. MSgt. Portan advised that the decision to retrofit the Thunderbirds flight helmets was directed by Air Combat Command.

Since the Thunderbirds did not stock Combat Edge kits, they had to reach out to AFE shops across USAF enterprise to find a minimum of 6 kits to upgrade the delta formation. Air Combat Command gave the AFE Shop 72 hours to comply, or switch to USAF issued grey, HGU-55/P flight helmets. This was a daunting task, but it was in the interest of pilot safety. And, no one wanted to switch to grey USAF issued HGU-55/P flight helmets. No one!

As kits began to arrive, new challenges were realized. The Combat Edge kit comes with a clear plastic template (photo on the right) that uses the visor snap and nape strap screw for proper alignment of the bladder inflation tube access port. The problem is, the Thunderbird helmets do not retain the visor snap, and upgraded the nape strap for the integrated chin and nape strap, which requires different mounting holes. Without these critical alignment references, the AFE specialists would have to "eyeball" the location of the template. Add to that the unnerving responsibility of drilling and filing the port opening and attachment screw holes through that beautiful paint job! This was not taken lightly!

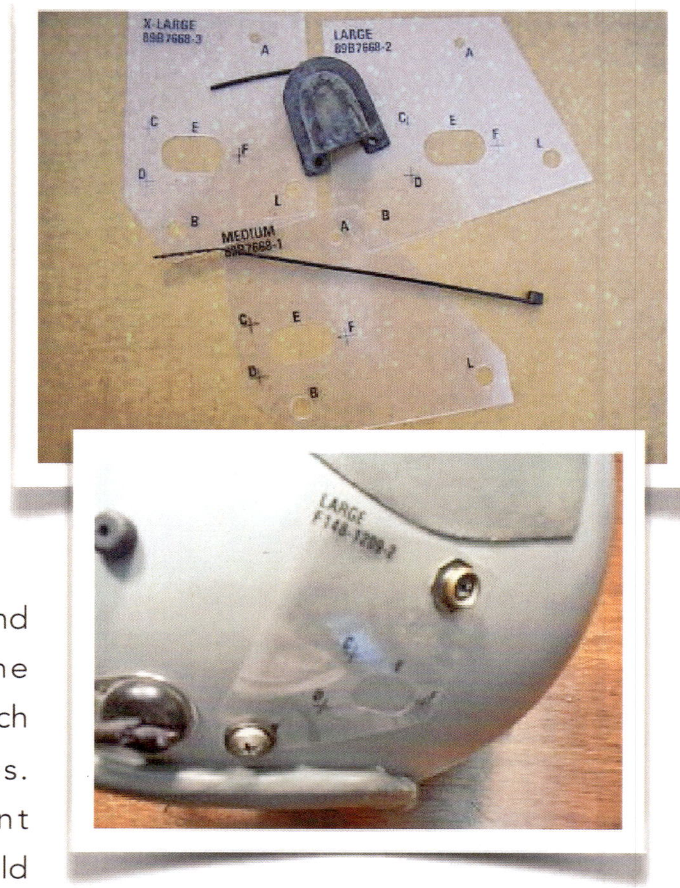

Photo credits: flighthelmet.com

TSgt. Jasper Roberts was tasked with drilling the first helmet. Tensions were high as everyone feared the paint would chip or crack; however, their fears were unwarranted. The paint and clear coat was resilient and the helmets were successfully upgraded without damaging the finish. Once again, mission accomplished.

The photos on this, and the following page are the flight helmets of the 2019 USAF Thunderbirds team.

Lt. Col. John "Brick" Caldwell

Maj. Will "Boar" Graeff

Capt. Michael "Thorny" Brewer

Maj. Whit “Skate” Collins

Maj. Matt “Spider” Kimmel

Maj. Michelle “Mace” Curran

United States Air Force Thunderbirds Flight Helmets

As my "time" came to a close with the Thunderbirds, I thought it would be appropriate to share a photograph of the clock in the AFE Shop. It is, well, what you might — or might not — expect…one half of a Thunderbird flight helmet and oxygen mask! Twelve officers, and 12 hours of the day. If you look closely at the numerals, the #5 is inverted as it appears on the Lead Solo jet, which spends much of its air show routine in inverted maneuvers. This gives me an idea, but I can't afford to cut a Thunderbird flight helmet in half!

15
The Quest

For as long as I can remember I've had a fascination, or obsession depending on your perspective, with aviation. I was only 16 years old when I learned to fly in 1974, and the following year, I witnessed the USAF Thunderbirds perform for the first time in my life.

I made the short 75 mile flight in a rented Cessna C-172 from Lafayette, LA to Lake Charles, LA (Chennault Air Base) in mid October for the 7th annual Young Men's Business Club (YMBC) air show. The Thunderbirds had just recently transitioned from the McDonnell Douglas F-4E "Phantom" to the Northrop T-38A "Talon" due to the 1973 fuel crisis.

The Thunderbirds performance was awe inspiring, like nothing I had ever witnessed, and it left a lasting impression on me. After 45 years since that first Thunderbirds air show, I still travel throughout the U.S. and make a concerted effort to see them as often as I am able.

United States Air Force Thunderbirds Flight Helmets

Another noteworthy air show I attended was 21 years later in 1996 (Lafayette Regional Airport, Louisiana) which marked a turning point from my interest in the Thunderbirds as a flight demonstration squadron, to the allure of their awesome flight helmets!

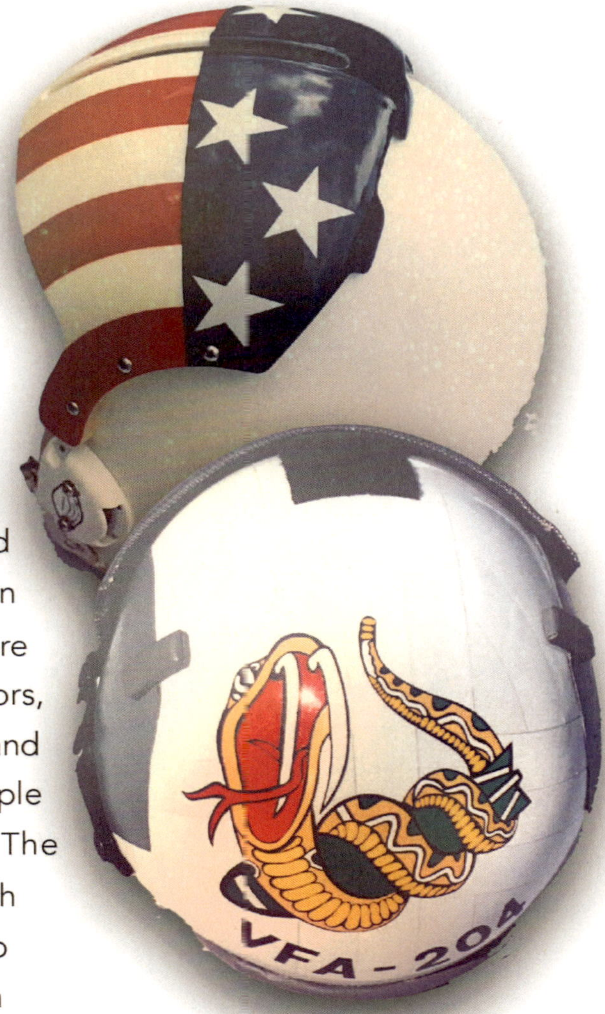

Author's custom flight helmet work, personal, private collection.

In the early 1990s, my flight gear collecting interests evolved into restoring military flight helmets. I would rescue old flight helmets sentenced for destruction by DRMO[1], disassemble all removable parts, sand, fill, prime and repaint the shells to "like-new" condition. I learned to paint from my father-in-law (an award winning professional automotive painter) and another good friend I knew from my days in law enforcement. These first helmets were pretty simple and straightforward, single colors, no graphics; however, I soon began to expand my horizons and experiment with multiple colors, patterns and reflective tape.[2] The helmets would be reassembled, often with new, old stock (NOS) parts, similar to restoring classic automobiles, but on a much smaller scale.

In that same year, I began to concentrate on my historical research of the USAF Thunderbirds flight helmets. In my ongoing search for technical and historical information, I made an untold number of "cold calls" to the

---

[1] DRMO - Defense Reutilization and Marketing Service manages the reuse, transfer, donation, sale or disposal of excess / surplus property.

[2] Reflective tape is used on U.S. Navy flight helmets to aid in the night recovery of a downed aviator at sea.

Thunderbirds and Flight Suits, Ltd., (FSL)[3] the single source provider of custom flight suits and flight helmets to the Team.

I learned that it was possible to have FSL make a Thunderbird flight helmet for customers willing to shell out about $700 - $800[4] for a high quality "replica." I also stumbled across an auction for an authentic Thunderbirds helmet (issued to, and flown by a team member) that fetched almost $2,000 from the successful bidder! I wanted one of these helmets, but I couldn't compete with those prices!

Intent on adding one of these helmets to my personal collection, I decided to recreate one. I felt I was up to the task, but the challenges that lied ahead would test not only my talent and ability, but my patience as well. I have been characterized as a perfectionist, defined as rarely satisfied with my work, by people who know me. I realize that perfection is unattainable so, I focused my efforts towards achieving satisfaction, despite anticipated imperfections.

With a lot of persistence and a little luck, I established a strong relationship with the FSL Thunderbirds Account Manager who turned out to be a wealth of information as well as kind, supportive and inspirational.

Author's custom flight helmet work, personal, private collection.

[3] Now Gibson & Barnes.

[4] In 1996 dollars.

Over the course of about a year, I was able to obtain authentic USAF Thunderbird flight helmet decals and graphics for my project. In addition to the decals, I was provided with hand written instructions on the proper application technique, which is quite interesting in and of itself.

The "decals" are screen printed on clear acetate with an adhesive backing. The instructions involved cutting the decal into several "sub-components" to enable the emblem to more easily conform to the compound curvature of the helmet. Following paint, decals and graphics, the helmets receive several applications of urethane-based clear coat followed by wet sanding, polishing and buffing. The results are truly stunning!

During this research period, I also made many phone calls to the Thunderbirds Life Support Shop, as well as arranged personal meetings with Life Support personnel at local air shows to answer additional technical questions, and to take detailed photos of actual team helmets as a point of reference. These would prove to be invaluable.

The "Quest" was about to begin.

After receiving my decals, graphics, instructions, paint and clear coat, the time had come to begin this arduous and daunting task. I had previously selected a "stripped" helmet shell — slated for destruction by DRMO — that I meticulously sanded, filled and primed. The primer coat was wet sanded with 600 grit automotive sandpaper to ensure the surface was smooth and free of any visible imperfections. The 600 grit was also a good choice to provide the right "tooth" to allow the paint to adhere to the surface of the helmet shell.

The paint choice (in the mid 1990s) was DuPont Imron[5], highly favored by the aircraft industry for its durability and high luster. It was expensive…very expensive but, I decided early on to do this right and not cut any corners.

[5] Currently using PPG paints.

Early measurements, calculations and line drawings on a "donor" shell in preparation of transferring to the painted helmet.

The artists at FSL suggested to paint the shell with ABS 100 ("Thunderbird Red") first, followed by wet sanding, mapping out the "scallops" and shooting the DuPont Imron 555 (white). I followed these instructions, but later learned it is easier to paint dark over light, rather than trying to cover dark paint with a lighter one. I would make this change on future helmet projects.

The application of "Thunderbird Red" was satisfying and without consequence. After about a 24-hour drying period, the red was carefully wet sanded with 600 grit sandpaper being very cautious not to sand through the delicate paint! The sanding removed slight surface imperfections (sometimes referred to as "orange peel") and all of the glossiness. The end result was a flat red, lusterless surface.

Final measurements transferred and taped in preparation for white paint.

Mapping out the scallops was without a doubt the most tedious and time consuming part of the project. Although the written instructions were very detailed, working on a compound curve comes with many challenges! I used another plain white helmet shell to work through the calculations necessary to "plot" the numerous points, spaces and radii that form the graceful scallops. Once satisfied with my math, I simply transferred this data onto the red shell. From here, it was an exercise in "connecting the dots" with 1/16-inch painters tape to establish a clean, accurate paint line (previous page).

As expected, the white required a few more coats than the red in order to achieve complete coverage. The anxiety came when it was time to remove the tape. If I removed it too soon, I risked the paint sticking to the tape like melted mozzarella cheese. If I removed it too late, the paint would dry on the tape and pull it off the surface when peeled away. In the end, "Lady Luck" was on my side, and I was rewarded with crisp paint lines.

As with the red paint, the white paint was also sanded with 600 grit sandpaper. Again, caution was the order of the day, being careful not to blow through the white paint, or, damage the fresh paint lines. Thanks in part to my experience from previous painting projects, I was able to avert disaster.

The application of graphics was the next order of business. The folks at FSL sent me some handy templates for the proper alignment of the Thunderbird emblem, which made the application "almost" foolproof. The "segmented" emblem was applied with a soapy water solution until proper alignment was achieved, then the excess water was blotted and "squeegeed" until set. Once the stars found their proper places relative to the scallops, the helmet was allowed to dry overnight.

The clear coat was a very rewarding process, bringing life, depth and sparkle to the newly laid graphics. The helmets I've seen from FSL have several coats of clear, but upon close examination, the graphics are slightly raised, or "proud" of the surface. I decided to apply enough clear to create a completely smooth surface. This required applying 4 coats of clear, allowing overnight dry time, and lightly wet sanding with 800 grit sandpaper, knocking down the high spots. This process was repeated 3 more times until 12 coats were applied, completely burying the graphics.

More wet sanding. I started with 800 grit until the surface was smooth and all of the sheen had been removed. Then I move on to 1,000 grit, 1,500 grit, 2,000 grit and finished with 3,000 grit sandpaper. The final sanding was so smooth, the luster was beginning to return. Next, I polished the clear coat with a high quality automotive rubbing compound and a power buffer. The final step is the application of a high quality automotive cleaner and wax. It wasn't perfect, but I was satisfied! Mission accomplished!

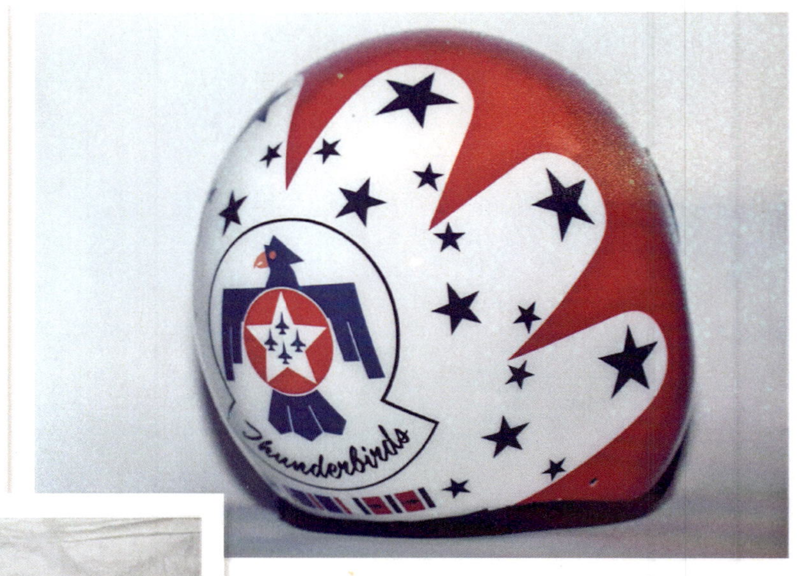

Once I had reached this step, it was time for a custom, hand sewn leather edge roll, a technical skill I didn't possess at the time, but learned several years later. I sent my helmet shell to FSL where they professionally installed a new leather edge roll and lined

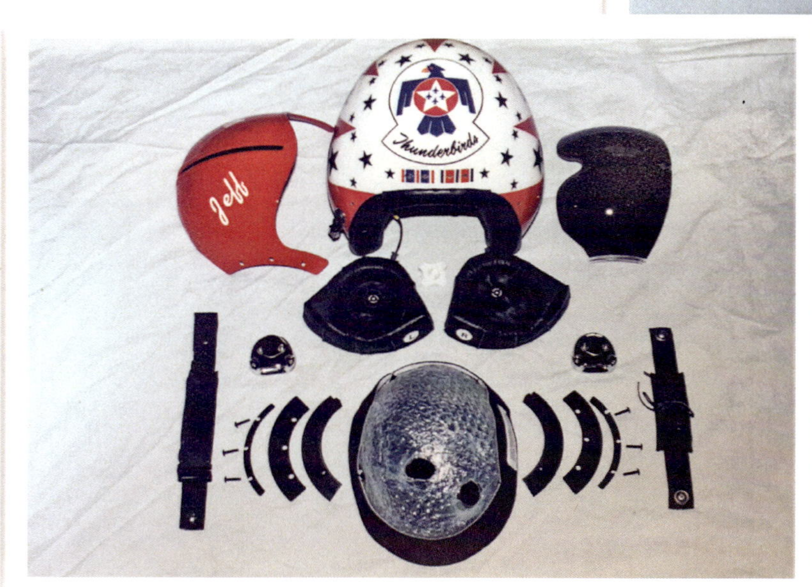

Graphics applied; Clear coat applied; Polished and buffed; New custom edge roll by Flight Suits, Ltd.

the interior with Tempo[6] fabric. When I received my helmet back from FSL, it came with the following compliment from the Thunderbirds Account Manager:

[6] Tempo is a U.S. Trade Marked textile that allows Velcro to adhere to it.

"We're very impressed — it is a work of art! We can tell you are a perfectionist. We'll never see another 'garage variety' Thunderbird helmet this good!"

One of the Thunderbird helmet artists at FSL also commented that the first helmet he made didn't look nearly as good as mine! This validation was quite an ego boost and provided me with additional motivation to pursue this passion-turned-obsession!

It was now time to put all the pieces back together: Instal the liner, earphones, communications cord, attach the visor, visor housing and visor lock knob, replace the chin and nape straps, and attach the freshly chromed oxygen mask bayonet receivers. At the end of the day, and a lot of hard work, it looked like a million bucks!

Several years passed, and after much internal debate, I decided to recreate a replica Thunderbird helmet of each different style used by the team since their inception in 1953. This not only stoked the creative side off my brain, but furthered my research on the history of these awesome helmets. I hope you enjoy this photo-essay!

ALL PHOTOS ON THE FOLLOWING PAGES ARE THE WORK AND INTERPRETATION OF, AND IN THE PRIVATE COLLECTION OF THE AUTHOR

(Missing from the author's collection, but in the project queue, is the HGU-55/P/E Combat Edge flight helmet - the most current helmet in use at this writing.)

P-1 Series Flight Helmet
F-84G "Thunderjet"

P-3 Series Flight Helmet
F-84F "Thunderstreak"

Mine Safety Appliance, TOPTEX
F-100C/D "Super Sabre"
F-105 "Thunderchief"

HGU-2A/P & HGU-26/P Series Flight Helmet F-4E "Phantom"

HGU-2A/P & HGU-26/P Series Flight Helmet
T-38A "Talon"

HGU-55/P/E Series Flight Helmet
F-16A/B/C/D "Fighting Falcon"

HGU-55/P/E Series Flight Helmet
Integrated Chin / Nape Strap
F-16C/D "Fighting Falcon

HGU-55/P/E Series Flight Helmet
Combat Edge
F-16C/D "Fighting Falcon

"Currently in the Project Queue"

16
The Future

While it is impossible to predict the future, the fact remains that the Thunderbirds have been flying the F-16 "Fighting Falcon" — a 4th Generation[1] aircraft — for more almost 35 years. This is the longest any aircraft has been in Thunderbirds service. Because these jets are far from new when the Team receives them, they are all stressed and tired airframes despite the meticulous maintenance they receive on a daily basis.

Many have speculated and many more have fantasized — myself included — that the Thunderbirds are considering upgrading their demonstration aircraft to a 5th Generation[2], front line USAF fighter. With only 2 aircraft currently in this category, the choices are limited to the F-22 "Raptor" or the F-35A "Lightning II" both manufactured by Lockheed Martin.

The highly advanced F-35 features a high-tech helmet capable of assisting the pilot to virtually see through the plane. When the pilots look down, they don't see the floor of the plane; they see the world below them. If the pilots look back, they see the sky behind them. Embedded in the skin of the aircraft are six cameras, and when the pilots move their heads to look in a particular direction, they are actually seeing through the corresponding camera, which sends an image to projectors inside the helmet that beam an image of the outside world on the helmet's visor.

The imagery is projected onto the pilot's visor, like a screen that posts information the way some cars are now posting fuel and gas mileage on the windshield in what's called a heads-up display (HUD). But beyond speed and altitude, F-35 pilots would see things such as the location of enemy aircraft or weapons on the ground.

---

[1] Aircraft in service from approximately 1980 to the present and represent design concepts of the 1970s.

[2] Aircraft featuring all-aspect stealth, low-probability-of-intercept radar (LPIR), high-performance airframes, advanced avionics, and highly integrated computer systems.

Like the plane, the F-35 helmet is enormously expensive. The cost of each custom-made helmet is more than $400,000[3] an example of which is shown below.

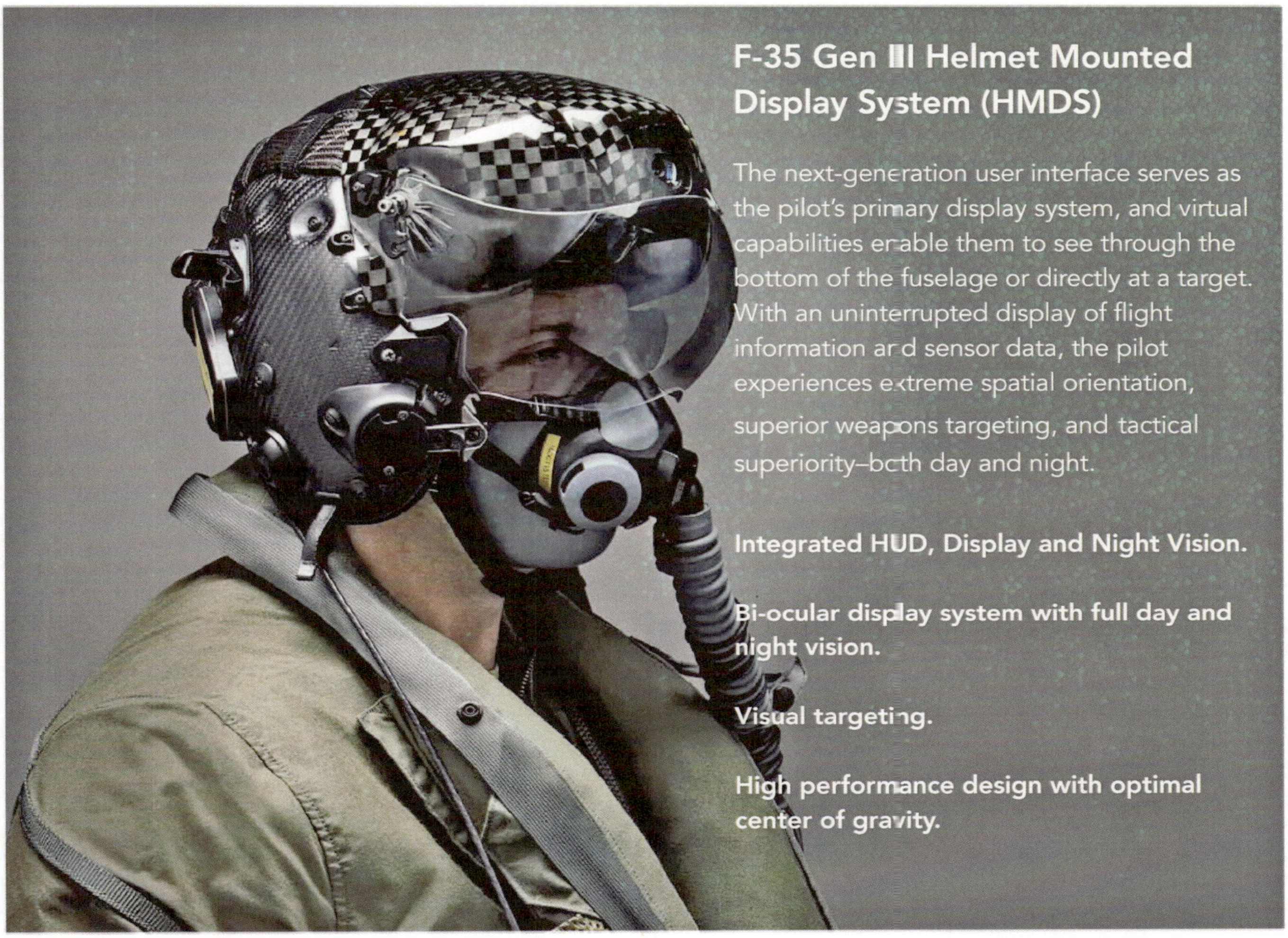

It may be argued that these helmets do not lend themselves well to the "blank canvasses" of P-series and HGU-series flight helmets used by the team in previous years and that would be a fair assessment. These are very sophisticated, high tech and complex flight helmets that are "tools" to assist the pilot in a combat environment. They are also larger and heavier than conventional flight helmets used in 4th Generation aircraft. Because the Thunderbirds would not have the need or desire to

[3] https://www.stripes.com/news/us/400k-helmet-helps-f-35-lightning-ii-pilot-see-through-the-plane-1.337891

access these combat-related systems while performing an aerial demonstration, one might suggest that the Thunderbirds would continue to use a variant of the HGU-55/P, wired specifically for either of these two aircraft.

If history repeats itself, the USAF Thunderbirds will not be satisfied continuing to fly a 4th generation fighter while new, shiny 5th generation aircraft are operational. With the exception of the T-38A "Talon" (due to the 1973 fuel crisis), the Thunderbirds have always flown front-line fighters. But this comes with a cost. These decisions are made at very high levels and contingent upon inventory and available budgets.

Speaking of budgets, a Block 50-52 F-16D costs only $18.8 million.[4] The F-22 per unit cost is +/- $150 million[5], while the F-35 per unit cost is +/- $91 million[6]. Multiply this times 6 show aircraft, the narrator jet, advance pilot jet and a couple of spares, and it's not long before you're talking about real money! And, this doesn't even consider maintenance!

Adding to this complex issue is another obstacle: The F-22 and F-35 are single seat aircraft only. Two seat variants were not developed in order to keep production costs down. This poses a problem for the Thunderbirds as they rely heavily on two seat model aircraft to teach new team members — including the commander / leader — their respective positions in the diamond and delta formations. This would also negate any media or VIP flights should the team choose either of these platforms.

A possible option would be to transition to the new Boeing T-7A "Red Hawk," a single engine, advanced jet trainer slated to replace the aging T-38 "Talon" fleet. According to Boeing, the maintenance-friendly design supports long-term functionality and allows for the integration of the latest technology and equipment. The aircraft serves as a more affordable and flexible platform when compared with the existing aircraft in its

---

4 https://www.af.mil/About-Us/Fact-Sheets/Display/Article/104505/f-16-fighting-falcon/

5 https://militarymachine.com/f-22-raptor-vs-f-35-lightning-ii/

6 https://militarymachine.com/f-22-raptor-vs-f-35-lightning-ii/

class. The system also supports technological evolutions to meet the requirements of future training needs.

Similar to the 1973 fuel crisis, this could be a modern day answer to a growing budget concern. It's not a 5th generation fighter, but it is a state of the art, sleek, sexy trainer, not unlike the T-38 was 45 years ago. I, for one, would not be disappointed!

This dilemma is also shared by the Thunderbirds' sister organization, the U. S. Navy Blue Angels. The F-35 is a multi-role, joint strike fighter in the inventories of 3 U. S. service branches: F-35A (USAF), F-35B (USMC) and F-35C (USN). Instead of upgrading to the F-35C, the Blue Angels have opted to trade in their McDonnell Douglas F/

A-18A/B/C/D "Hornet" aircraft for the Boeing F/A-18E "Super Hornet" — a formidable, but 4th generation fighter[7].

At the end of the day, these issues will be properly sorted and decisions will be made, by the right people, at the appropriate time and appropriate place. One thing is certain: The Thunderbirds' mission remains strong, evidenced by their primary objectives:

1. Support Air Force recruiting and retention programs.
2. Reinforce public confidence in the Air Force and to demonstrate to the public the professional competence of Air Force members.
3. Strengthen morale and esprit de corps among Air Force members.
4. Support Air Force community relations and people-to-people programs.
5. Represent the United States and its armed forces to foreign nations and project international goodwill.

These five objectives are deeply rooted in the team's heritage and essential to the mission of the United States Air Force. Regardless of which aircraft is selected to replace the F-16 or their choice of flight helmet, we will always look forward to the *next* season in anticipation of what the USAF Thunderbirds have to offer an eager and growing air show audience!

Blue Skies!

[7] Tentative planning underway in 2018.

APPENDIX

USAF Thunderbirds Decorations

The respective award dates can be a useful tool in determining the age or era of a flight helmet.

Air Force Outstanding Unit Award
1958 (1st)

Air Force Outstanding Unit Award
25 FEB - 31 DEC 1968 (2nd)

Air Force Outstanding Unit Award
1 JAN - 31 DEC 1973 (3rd)

Air Force Outstanding Unit Award
1 JAN - 31 DEC 1974 (4th)

Air Force Outstanding Unit Award
1 JAN 1979 - 31 DEC 1980 (5th)

Air Force Organizational Excellence Award
1 JAN 1984 - 31 DEC 1985 (1st)

Air Force Organizational Excellence Award
1 JAN 1986 - 31 DEC 1987 (2nd)

Air Force Organizational Excellence Award
30 SEP 1989 - 30 SEP 1991 (3rd)

Air Force Outstanding Unit Award
1 JUN 1995 - 31 MAY 1997 (6th)

Air Force Organizational Excellence Award
1 JUN 1997 - 31 MAY 1998 (4th)

Air Force Outstanding Unit Award
1 JUN 2001 - 1 JUN 2003 (7th)

Air Force Outstanding Unit Award
1 JUN 2004 - 31 MAY 2006 (8th)

Air Force Outstanding Unit Award
1 JUN 2006 - 31 MAY 2008 (9th)

Air Force Outstanding Unit Award
1 JUN 2008 - 31 MAY 2010 (10th)

Air Force Outstanding Unit Award
1 JUN 2010 - 31 MAY 2012 (11th)

Air Force Outstanding Unit Award
1 JUN 2012 - 31 MAY 2013 (12th)

Air Force Outstanding Unit Award
1 JUN 2013 - 31 MAY 2015 (13th)

Air Force Outstanding Unit Award
1 JUN 2015 - 31 MAY 2017 (14th)

(NOTES: 1. There are no known examples of Thunderbirds flight helmets with two Air Force Outstanding Unit Award ribbons, denoting the 10th and 14th awards; 2. The inclusion of award ribbons below the Thunderbirds emblem was discontinued around 2015.)

Index

Acknowledgements

It would have been impossible to take on this monumental task of research, documentation and editing without the support, encouragement and patience of many people. First and foremost, I would like to thank my wife, Renée for her patience and support during those long hours of research, road trips and endless hours on the computer. I would also like to thank my editor, Ed Roy, and my publisher, Mick Prodger of Elm Grove Publishing for his faith in me as an unpublished author and his expertise in guiding me through this arduous project. I owe a debt of gratitude to the following people for their selfless time and talent by providing me with interviews, facilitating tours and sharing technical data. Without them, this book would not have been possible.

The Major Stephen "Cajun" Del Bagno Family

The United States Air Force Thunderbirds Public Affairs

The United States Air Force Thunderbirds Aircrew Flight Equipment Shop

The Department of the Air Force, Public Affairs & National Media Engagement Office

Thunderbird Alumni Association

Lieutenant General C. C. "Buck" Pattillo, USAF (Deceased)

Brigadier General E. Dan "Boss" Cherry, USAF (Retired)

Brigadier General James Latham, USAF (Retired)

Colonel Mike "MO" Beale, USAF (Retired)

Colonel Ronald Catton, USAF (Retired)

Colonel Eric Schnaible, USAF (Retired)

Colonel Charles C. Pattillo, Jr., USAF (Retired)

Lieutenant Colonel Caroline "Blaze" Jensen

Major Ray Geoffroy, USAF Thunderbirds Public Affairs

Major Nate "Hassel" Hofmann, USAF

Major William "Boar" Graeff, USAF

Major Michelle "Mace" Curran, USAF

Major Trevor "Dozen" Aldridge

Master Sergeant Christine E. Powers, USAF (Retired) Thunderbirds Public Affairs

Master Sergeant Edwin Portan, Aircrew Flight Equipment

Staff Sergeant Ned Johnston, USAF Thunderbirds Public Affairs

Mark Jones, GENTEX Corporation

James Wegge, Gibson & Barnes

www.ingramcontent.com/pod-product-compliance
Lightning Source LLC
LaVergne TN
LVRC101545220926
848486LV00041B/383

* 9 7 8 1 9 4 3 4 9 2 7 6 3 *